DESIGN AND INNOVATION

DESIGN AND INNOVATION
POLICY AND MANAGEMENT

Edited by

Richard Langdon and Roy Rothwell

St. Martin's Press, New York

First published in the United States of America in 1985

ISBN 0-312-19448-X

Library of Congress Catalog Card Number: 8526109

Printed by SRP Ltd, Exeter

CONTENTS

Acknowledgements

The work that led to the production of this book was funded by the Wolfson Foundation, and we are grateful to Dr Black, the Director, and to the Trustees of the Foundation for their support.

I am indebted to Thierry Gaudin for agreeing to be the second Visiting Wolfson Professor of Design Management, and for his contribution to the workshop.

I am also grateful to Charlotte Coudrille, whose patience, expertise and skill in sub-editing and production were invaluable in preparing the work for publication.

Richard Langdon
London 1985

FOREWORD

Richard Langdon

Thierry Gaudin was the Visiting Wolfson Professor of Design Management at the Royal College of Art during 1984, and he chaired the workshop entitled *Design and Innovation: Policy and Management*. He played an important role in setting up the workshop and contributed significantly to the discussions during the sessions. His main contribution was to describe the work of his group in the Ministry of Industry and Research. This group has had a significant impact on French government policy and the implementation and management of that policy. The primary aim of French policy has been to encourage technological and design developments in selected industries, in particular by using government procurement as a means of stimulating design and innovation. Gaudin gave a number of examples including that of the furniture industry, where the refurnishing of government offices has been undertaken in a way that will encourage a revival in the industry.

Professor Gaudin is an acknowledged international expert in innovation policy and is past chairman of the Six Countries Programme on Aspects of Government Policies Towards Innovation in Industry.

The purpose of the Centre de Prospective et d'Evaluation (CPE), which Professor Gaudin directs, is to promote forecasting and evaluation at the forefront of technology and industrial developments, in collaboration with other leading French policymakers. Forecasting and evaluation can not be separated: all forecasting is based on an idea of a desired future; all evaluation refers to the value which various actions create. The main idea of CPE is that the combination of forecasting and evaluation can transform social practices. More precisely, CPE seeks to ensure the development of coherent visions of the future. It surveys foreign technology and science, diffuses the results, and evaluates their impact in France. In addition, it uses and publicizes methods of evaluating social systems: research establishments, aid programmes, technological programmes, administrative structures, nationalized enterprises, and so on. It also supports research into research, the relationship between science, technology and society, social evaluation of technology, international comparisons of industrial policies and policies of research and development. It organizes the annual national "Inova" exhibition.

CPE brings diverse expertise together in a modest establishment, and also cooperates with numerous private and state institutions. Its task is to translate into action existing research and social practice, thought and techniques, and it carries out its own research for this purpose. In addition it responds to the Ministry's needs for analyses bearing both on strategy and implementation. CPE shares its site at Montagne Sainte-Genevieve with "Cesta", the Centre for Studies of Systems and Advanced Technologies, a public research establishment.

INTRODUCTION

Richard Langdon and Roy Rothwell

In his lecture on "Design and British Economic Performance", Freeman proposed a much wider range of scientific and technical activities, which are important for the efficient conduct of technical change and for a high level of economic performance and which are not covered by the present definition of research and development. He called this *Research, Design and Development*. He points to the principal elements in the long-term techno-economic strategy and illustrates this in the context of different national policies directed towards innovation. In the case of Japan, a systems approach to design was adopted which recognized its integrative role in the management of innovation; a national-level strategy was created to bring together the best resources from universities, government research and private industry to work on important research, as well as on design and development problems; and lastly, a national policy was adopted toward R,D&D strategies, including investment and training.

In *The Evolutionary Theory of Economic Change*, Nelson suggests that the term "innovation" implies a change in routine, and further that there is considerable uncertainty surrounding technical innovation, which is the implementation of a design for a new product or of a new way to produce a product. A similar uncertainty surrounds other kinds of innovation − the establishment of a new marketing policy, or a new decision rule for restocking inventories.

Schumpeter identified innovation with the "carrying out of new combinations". Innovation in the economic system consists of the recombination of conceptual and physical materials that were previously in existence. The vast momentum of scientific, technological and economic progress derives not merely from the solution to a particular problem, but adds a new item to the set of components which are then available for recombination in the solutions to other problems in the future. Simon, in *The Sciences of the Artificial*, in comparing design to science, says: "Design on the other hand is concerned with how things ought to be, with devising artefacts to attain goals."

The greatest stimulus to contemporary work in design methods has been the combination of interest in the design of large and complex systems and in the growth of knowledge in computing and information technology. The requirements of society for power systems, transportation systems, data processing computer systems and automatic control systems is indicative of the size and complexity of this growing problem. In recent years the efforts to study design and development activity has produced a greater understanding of the design process. The formalization of methodologies has led to the development of a conceptual and theoretical framework for design. The dominant view of design theory holds that design is a cognitively based, problem-solving activity.

Design method describes and models the design process as cyclical since, in most cases, it passes through a number of iterations before a solution is generated. The purpose of design is to create or restructure a situation-specific component, product or service which effectively achieves engineering goals and meets organizational or social needs. So that to conceive the idea for, and prepare a description of, a proposed system or artifact, whether this is an automobile, an electronic circuit, a management system or an economic model, involves processes, steps and knowledge which are common to all domains of application. Design is a process in which a concept describable in language, symbols and figures is transformed into an entity that can be assessed through its attributes by scientific means. This entity must satisfy requirements described in the concept and be complete enough for production. Design is a creative process by which products and processes are conceptualized and specified, and as such has a vital role to play in enabling companies successfully to exploit their innovative research.

The R&D decision maker does not know the economic characteristics of as yet unconfigured technologies, but he knows certain technological attributes. He will have an idea of what sub-classes of new technology will lead to a product with one or other set of attributes. A good decision rule for allocating R&D funds must attend both to factors on the demand side and to factors that influence the ease or cost of implementing invention. The new is not just better than the old, it evolves out of the old. The output is not merely a new technology, but also enhances knowledge in many hardware industries. R,D&D can be represented as a gradual filling in of details in the overall concept, the course of the design work being guided by a series of studies and tests. In the later stages these involve prototype versions of actual new hardware. R,D&D is building a technological variant that was not in existence before, and finding out how it works. Information is being acquired not only in activities that are incidental to discovery, but also in the course of creating and learning about something new. In general the new design (configuration) will involve a large number of subdesign elements or components. Regarding each of these there may be certain design problems to solve, in the sense that certain performance goals (criteria) need to be met. Knowledge can facilitate the problem solving by guiding the effort toward promising design alternatives (different configurations).

The current configuration represents a set of solutions to design problems and provides the starting point for the next round of R,D&D. This formulation appears to explain what is known as the product cycle. A product lasts only for a limited period of time, which creates pressure on the company to develop new products. This cannot be done at random and requires the formulation of a policy for the innovation process. Within the company this is often called corporate strategy. The study of these patterns of product development has proved to be of great importance in providing knowledge on which to base future decision making, both in terms of policy making and the subsequent management and handling of uncertainty. The lack of case studies and project histories has hampered much of the policy making in the UK, where neither civil nor military projects are studied to the same extent as, for example, in the United States and Japan. Thus the meta-level knowledge is not made extant and cannot therefore systematically become part of policy making. This has led to a largely arbitrary approach to publicly funded new product ventures and, given the greater uncertainty and high risk, has led not surprisingly to a number of projects which have been unsuccessful. This problem has been recognized in Japan, where MITI has adopted an active policy towards identifying and collecting information and know-how about high technology and rating the knowledge. In a recent MITI report on different manufacturing sectors in product and production technology, it has been identified where resources should be placed. In 1981 MITI announced a programme known as "Next Generation Industrial Foundation Technology Development System". It was based upon the assessment that Japan has to concentrate her R,D&D efforts on basic technologies where in many cases they were ahead of Western countries, who would in any case become increasingly reluctant to allow the transfer of technology. The report indicated a remarkable conformity of aims between MITI and Japanese companies in their trade aims.

New technologies, of course, do not always offer immediate benefits over existing systems. In many cases the first versions of the new technology tend to be only marginally superior, and sometimes not superior at all to existing systems. The advantages achieved were largely through the wave of improvements made possible by the new design, compared with the difficulty of finding further major improvements in the old one. The concept of "robustness", proposed by Rothwell and Gardiner in relation to product design and development, is a useful one and in the context of cybernetics it is known as the concept of requisite variety. In applying this concept, there should be built into any manufactured product or system a quantity of variety so that it is capable of accommodating itself to initially unknowable future changes in production, markets and environments in which it is used. The internal connectivities of a product or process should be sufficiently rich and the tolerances on its internal and external connectivities should be robust, so that the product or process can be repeatedly modified and restabilized. In the context of planning, robustness has been developed as a measure of the useful flexibility preserved by early decisions in a decision sequence. As the product evolves, so do the processes of production. Studies of the learning curves fall into three parts: workers learning to do the job better; management learning how to organize more effectively; and engineers redesigning the product and process to make it easier and cheaper to produce. The chapter Learning by Using in Rosenberg's book *Inside the Black Box: Technology and Economics* illustrates this concept with the example of aircraft designers who recognize that, although they configure the fuselage to conform to current capacity of the engines, it is generally understood that engines with improved performance will appear within the lifetime of the model, so that allowance for stretching is an important feature to be designed into the current configuration.

It was in order to tackle the above (and related) issues that a Workshop was arranged by Richard Langdon and Roy Rothwell at the Department of Design Research, Royal College of Art, in April 1984. The Workshop brought together an international group of experts and industrial practitioners in the broad fields of design and innovation management and public policymaking with respect to stimulating design and innovatory activities in industry. The primary output from the Workshop was a collection of papes written by a number of the participating experts. These have been edited and brought together in this book; they effectively survey the state of the art in the field of design and innovation policies and strategies and their implementation at both the public and corporate levels.

In the first chapter, Gregory tackles the issue of design strategy at the level of the firm. Since the product is the focus of business strategy it is, of necessity, the "valued object" towards which design is directed. In the past there has been relatively little obvious appreciation of the range of possible relationships between business strategies and design strategies. Since, however, business strategies have changed from an emphasis on integration, acquisition and disposal towards an emphasis on competitive activity and survival, the aims of business strategy and design strategy are rapidly converging.

Gregory describes a number of ways of approaching design strategy formulation and highlights the importance of taking into account "customer emphasis", "competition emphasis" and "company emphasis". He underlines the necessity for being aware of the complete span of objectives which have to be achieved by the valued object. Finally, Gregory emphasizes the importance of strategies towards designing for product differentiation; these, he contends, have been neglected in the past. Moreover, in approaching the issue of product differentiation, particular emphasis shouldbe placed on value creation as opposed to value analysis.

Nyström addresses the issue of company strategies for designing and marketing new products in the electrotechnical industry, an area in which design and innovation are critical factors determining competitive success. This feature, coupled with high rates of technological change in the industry, means that speed and flexibility in designing and marketing new products are of vital importance to success. This might, in certain cases, involve the need for entering contractual sales commitments for specified products before even prototype models are available, which in turn implies the need for high in-house technical competence. Under these conditions, an external R&D orientation, i.e. undertaking joint product developments with customers, becomes an attractive component of development strategy.

Attaining internal synergy between different developments and between functionally separated departments is also strategically important. In this respect, the level of integration between the technical and marketing elements can be greatly facilitated through flexibility in product design. A strategy of customization through the combination of different arrangements of standardized components greatly enhances product flexibility and functional variability.

Svidén, taking a futuristic view, looks at the possible influences of improved information technologies on the design and patterns of usage of the automobile. He describes a number of possible future scenarios: a "crisis" situation during the latter part of the 1980s, limiting energy supply to transportation; the year 2000, a society with high economic growth and a vital restructuring of industry in which new ecological and information technologies are well developed; 2010, an information society in transition; and 2040, a mature information society with decentralized organization and dispersed living.

The utilization of advanced information technologies in automobile design means that the car of the future will be more energy-efficient and less polluting, enjoy greater reliability, incorporate a wide range of automated functions, be safer, be more durable, incorporate systems for route planning, and be more flexible in its pattern of usage, i.e. will be multi-purpose in operation.

Kaplinsky discusses the issue of achieving comparative advantage through design. He plots a number of changes in the global manufacturing sector over the past forty years, emphasizing in particular the transition from "machinofacture" to "systemofacture", the latter owing much to the rapid diffusion of electronics-based automation technologies. Three different types of automation are categorized: intra-activity automation, which refers to automation that occurs within a particular activity; intra-sphere automation, which refers to automation technologies that have links with other activities within the same sphere; and inter-sphere automation, the most complete form of automation, which involves coordination between activities in different spheres of production.

The diffusion of the new electronics-based automation technologies has an important bearing on comparative advantage between the less developed countries and the advanced market economies in two different ways. The diffusion of intra-activity automation to the LDCs would have the effect of a continual enhancement of their comparative advantage, since it effectively reduces the need for skilled workers of the type in which LDCs are deficient. On the other hand, the more advanced inter-sphere automation is liable to find its greatest use in the advanced market economies, thereby enabling them better to withstand competition from the LDCs. In other words, the end result of technology diffusion could, in the long run, leave the pattern of comparative advantage largely unchanged.

One of the key issues covered by the seminar was the role of formal public policies in stimulating design and innovation activity at the level of the firm, and, as we shall see in the book, opinions on this issue varied considerably. Roessner, for example, expresses doubts as to whether *formal* public policies would be acceptable in the United States. According to Roessner, formal strategic and systematic design innovation policies would be acceptable in the United States only during periods of deep and prolonged crisis. The reasons for this are deeply rooted in US political values and manifested in its political institutions and policy-making processes.

Roessner does not suggest that successive US administrations have remained altogether aloof from industrial design/innovation activities, but rather that public policies have in the main been indirect rather than direct or specific rather than comprehensive. In other words, while the United States has never had strategic, comprehensive design/innovation policies, it has nevertheless promulgated a variety of taxation and economic measures and subsidies to tackle specific problems (inflation), to assist specific sectors of industry (civil aviation), or particular industrial interest groups (small businesses), or they have been designed to create an overall environment conducive to firm-based innovatory activities. These policies have not been indicative or coherent in the sense that they are associated with an overall National Plan.

Mowery and Rosenberg describe in detail an example of a targetted policy in the United States, that directed towards stimulating the technological development of the US civil commercial aircraft industry. They suggest that "The US commercial aircraft industry has been a major beneficiary of government policies that (prior to 1978) simultaneously affected the demand for and the supply of technological innovation." In this respect, they contend that US policies towards the civil aircraft industry resemble in many ways the policies adopted in Japan towards the computer and semiconductor industries. Moreover, according to Mowery and Rosenberg, this was a unique situation in the US context which occurred because of the close technological and financial links between the military and commercial aircraft industries.

Features of US policy that had a marked impact on the technological development of the civil aircraft industry are Federal regulations on licensing and the assignment of patent rights, the funding of inter-firm R&D activities, and the support of infrastructural R&D (in NASA) directed towards the civilian sector. Federal military procurement and research sponsorship created considerable civilian spin-off and greatly enhanced and expanded the pool of technological know-how available for application to the civilian aircraft sector.

A second example of specifically targetted policies in the United States is given by Colton et al. These are the National Science Foundation's Industry-University Cooperative Research Program and its Small Business Innovation Research Program. The first of these is directed towards the establishment of basic or applied research centres performing research of interest to both industry and academe. The second is directed towards stimulating new product development in small firms. These programs are small but effective.

In contrast with the United States where, for socio-political reasons, formal, indicative public policies are largely precluded, in Japan such policies represent the norm. In his paper on industrial policy, Kobayashi describes both the aims and means of industrial policies in Japan. The aims of Japanese policies are twofold: the first is to facilitate technological developments; the second is to encourage industrial structural adaptation. To achieve these aims requires that both government and industry adopt a long-term view and act in a coordinated manner. To this end, government must take a leading role in helping to determine future directions for industrial development and, through consultation with industry, create an atmosphere of cooperation in which to promote a unified "vision" of the future and decide on specific R&D targets.

Kobayashi provides an example of policy coordination and implmentation in the development of the Japanese computer industry. Key features of this programme are subsidized R&D, inter-firm cooperation in pre-competitive research and development and keen inter-firm rivalry in the marketplace following individual product development in the various participating companies. The current phase of the scheme is directed towards the development of the fifth generation computer. It involved establishing a new Institute for New Generation Computer Technology in 1982, staffed largely by engineers seconded from the eight participating companies, which is one hundred per cent government funded.

Ormala focuses at the level of a specific policy tool or institution and describes the role of the Electronics Laboratory of the Finnish Technical Research Centre as an innovation policy instrument.

Ormala begins by pointing to the lack of knowledge concerning the instrumentality of innovation policy tools due to the paucity of systematic evaluations of policy effectiveness. His paper represents a brief summary of the results of an evaluation of the research performance of the Electronics Laboratory and includes the importance of spatial factors and a determination of the impact and relevance of the laboratory's research projects.

Granstrand and Sigurdsen adopt a more global view and describe how policies in one group of countries have implications elsewhere. Specifically, they describe innovation policies in East Asia and their implication for Western Europe. They show how developments, particularly in Japan, have influenced innovation policies in Europe which has adopted a Japanese-style policy instrument, notably the pre-competitive collaborative research approach, as evidenced in the EEC-wide ESPRIT programme and the Alvey Programme in the UK.

Aubert, adopting a somewhat philosophical approach, discusses concepts of innovation policy and approaches to design. He describes the evolution of innovation policy from aids to development and implementing the application of research results, through the "institutionalization" of innovation policy phase to the present phase of managing technological and social change as a whole. He then discusses design in the "post-industrial" society and the role of the "scientific" designer and the "info-culture" designer.

Ashford tackles a specific policy area and discusses the instrumentality of government regulation as a stimulus for technological change. Regulation, Ashford contends, can be used effectively not only to achieve economic efficiency (which is where traditional economic regulation focuses) and to meet a variety of social and economic demands, but also to provide a stimulus for dynamic change. He provides a number of examples of regulation inducing technological change — not only compliance technology, but also ancillary innovations with commercial potential.

These empirical findings do, according to Ashford, have implications for design: the design both of regulatory systems and, in turn, of industrial innovation. He contends that regulations can be deliberately designed to yield a particular kind of technological response. Because of this, regulatory policy should be considered as an important instrument of innovation policy. To be effective this means, of course, a high degree of coordination between regulatory agencies and other public sector bodies dealing more directly with the stimulation of industrial technological change.

Casey attempts to link innovation theory to innovation policy. He describes three main developments in the theory and analysis of technological change: technology as a competitive factor of firms; the firm as the creator and evaluator of technology; and the relevant structural definitions of industry and the role of technology in cyclical and structural economic change.

In relation to the first strand of theory, policy should be based on an assessment of the requirements of different firms in specific industries, i.e. policy must have a strong sectoral basis. Moreover, the analysis of technology on its own will not suffice, and policy must take into account a variety of other factors, including the overall strategy of the target firm. Regarding the second strand of theory, appropriability of benefit is an important factor determining technology choice in firms, and there may be considerable differences between the objectives and responsibilities of governments and firms. For example, firms are concerned with the production of commercially viable technology, while governments might place considerable emphasis on the development of socially useful technology. The third strand of theory implies the need to understand a variety of inter-firm articulations. For example, there may be strong technological dependencies between groups of firms in different sectors. As a result, government policy should be concerned with stimulating and strengthening the linkages between technologically interdependent firms and sectors.

In the final chapter, Rothwell poses the question: "Are public innovation policies really necessary?" Evidence from the United States, where formal directed innovation policies are the exception rather than the rule, might be taken to suggest that the answer to this question is "no". Evidence from Japan, in contrast, suggests that innovation policies are not only necessary, but also that they can work spectacularly well.

However, the above difference between the United States and Japan might not, in practice, be as great as it would at first sight appear. Rothwell argues persuasively that while in Japan public innovation policies are both formal and *overt*, in the United States they are informal and *covert*. In supporting this argument he points to the crucial role played by NASA and the US Department of Defense in the creation of new, technology-based sectors such as satellite communications and semiconductors. More recent evidence suggests that the DoD's Strategic Computing and Survivability Program is, in effect, the US response to the Japanese government-supported Fifth Generation Computer Programme.

Rothwell concludes that innovation policymaking is far from being simply an economic and technological process; on the contrary, it is a political and cultural process. As a result, policy approaches which succeed in one country may stand little chance of success in other politically and socially quite different countries. Despite these differences, it is today an indispensable task of governments to stimulate increased national rates of market-oriented technological change. Well conceived and managed innovation policies are an important means to this end.

STRATEGY & DESIGN: A MICRO LEVEL VIEW

S.A. Gregory

There have been some substantial reports on design and its significance for the UK economy. These have been aimed nominally at individual companies and managers but, to a considerable degree, they have dealt with design and the economy at the macro level. There has been a considerable gap between recommendations in principle and the kind of activity which might be undertaken in specific companies.

Freeman[1] has gone into much detail regarding overall economic performance and its relationship to design. Rothwell, Gardiner and Schott[2] have made a set of observations which suggest kinds of activities in principle which might be undertaken by companies involved in design. Among other authors, e.g. Corfield,[3] and Walsh and Roy,[4] there have been recommendations regarding the application of design, and the representation of design at high level, but these have not gone much further. Corfield offered a design activity scheme which indicated useful points at which people other than designers might play a useful part (Figure 1). This type of scheme offers a formalization which is similar to the development of interfaces suggested by Flurscheim,[5] whereby useful relationships are ensured between functional departments and design work.

Recent authors on design management have tended to highlight a limited range of bureaucratic procedures in order to keep design work "on the rails". Very little of a positive nature has emerged.

Arising from experience, Gregory has reflected upon the apparent unwillingness of many managements to utter clear principles or directives. De Woot[6] has recently commented upon the extent of development in European management practice and noted the difficulty experienced in getting clear statements of strategic objectives. It is almost as if designers have to take part in an internal political process in which possible proposals have to be tried out on top management and others to see whether they strike fire. There is a second phase which is rather similar in the responsibility

<u>Company board</u> required to:

1 designate member with prime
   design responsibility

2 specify and obtain involvement of marketing,
   production and finance

3 establish a design review
   process

4 use product planning as
   base of investment
   decision

5 assure finance for full
   consideration of design and
   CAD in product development

<u>Before any development</u>, a responsible
check on:

a what market

b what performance to be attained

c correctness of price and revenue

d economy of manufacture

e use of up-to-date technology

f availability of materials and
   components

g reliability and maintainability
   relation

h identification of critical
   development phases

i achievability of delivery dates

PRODUCT PROPOSAL ENTRY

<u>Step</u>	<u>Action by</u>	<u>Action</u>
1	M, E	Identify need or want
2	M, E	Specification
3	M, P, E, F, L	Product relevance ***
4	E	Conceptual design
5	P	Preliminary cost
6	F, M	Evaluation **
7	E	Detail design
8	E	Prototype
9	P	Manufacture
10	M	Product launch
11	M, E, P	Product review **

Figure 1   Market-oriented design management procedure
            outline (essentially consumer goods)

it throws upon the designer. In the absence of clear directives there is no satisfactory basis for evaluation of proposals. Gregory has suggested that the designer should work by an ethical code which demands that, whether top management is intellectually committed or not, all the necessary features of a design, particularly those beyond narrow technical considerations, should receive adequate emphasis. Such a code demands interdisciplinary skills.

In discussing these proposals in industry, it has been suggested by some that lack of definition by management is almost a necessity in order to provide creativity. The writer views this without any sympathy. In order to be creative there have to be some ground rules; otherwise the problems become almost impossible to deal with.

Business strategies

Over the past two decades there has been a rising interest in business strategy. In the first phase there was an emphasis upon what to do in dealing with merger or diversification proposals. The area of interest today is with survival and growth within a competitive environment. Under these requirements, business strategy assumes more of the characteristics of military strategy. In this case the marketplace is the analogue of the battlefield, and the weapons used are the products devised for the market struggle.

Under such conditions, in which the individual company attempts to survive in a hostile environment, it is possible to invoke the principle of requisite variety, as set out by Ashby[7] for open systems. In order to remain viable, the system has to be able to respond to environmental threats and changes by generating internally suitable countermeasures. A countermeasure may be a change in management method or it may be a change in product. It is at this point that design and the ability to generate new design become perceived as part of the way in which the company can cope with its environment.

Practical business studies, such as those by Lawrence and Lorach[8] and later by Bessant,[9] have shown that, in addition to specialist responses, it is necessary to have some power of integration within the company (Figure 2).

The overall way of responding to the environment comprises the strategy of the company. This is developed through sub-strategies and tactics. New products or major product changes are a primary means of bringing the overall strategy into effect (Figure 3).

Johnson and Scholes[10] have provided an outline scheme of the kinds of activity which have to be undertaken in the process of

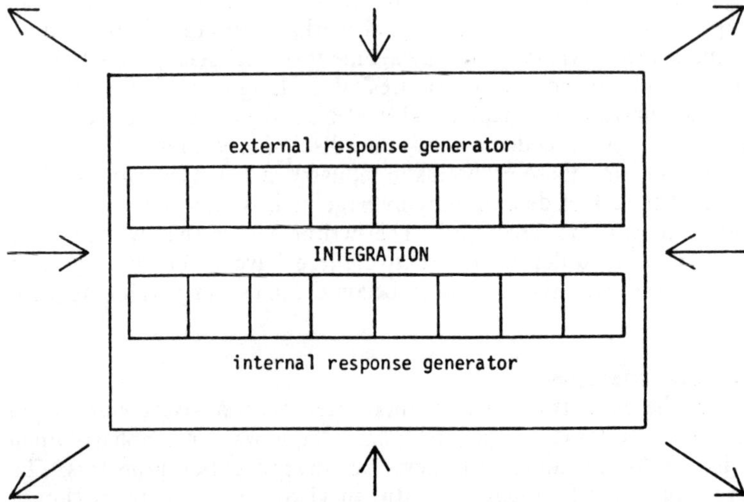

Figure 2   Open system in competitive environment
with generation of responses in integrated manner

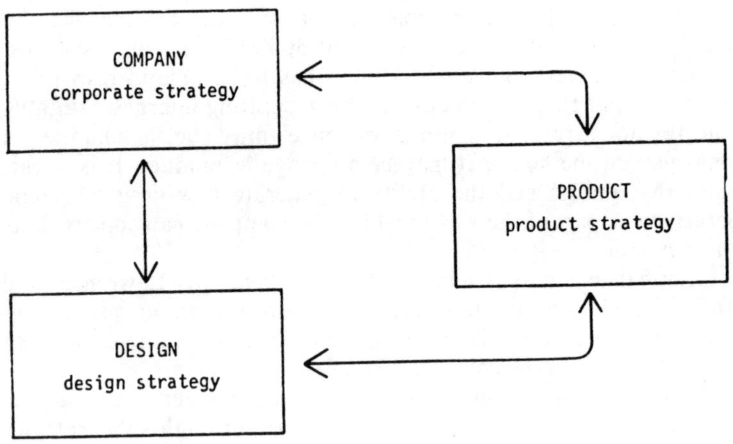

Figure 3   Relationships between company/product/design strategies

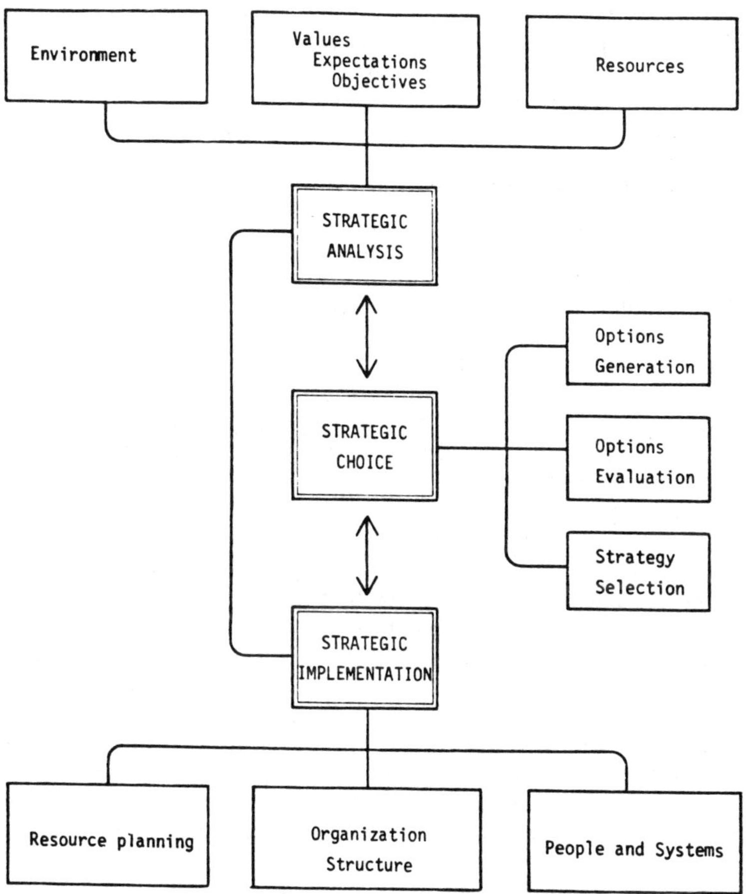

Figure 4   Corporate strategy development pattern

strategic activity, as depicted in Figure 4. Designers will note with interest that a strategy is designed and implemented. Provided that managers accept such a picture of strategic activities, they should also be able to grasp what design is about.

The work of Porter[11] has emphasized that there are generic strategies in business competition. In other words there are only a few basic patterns of behaviour which are invoked and their application turns upon the state of development of an industry, the phase in the relevant product life-cycles, and the structure of the industry concerned. In a broad sense a company has the following options in a market:

- cost leadership
- primary innovation
- product differentiation
- withdrawal

Positive options are elaborated in Figure 5.

The entry of design into strategy

Directly design becomes the means by which the strategic shift in product is made, it can be said that design itself has become strategic. However, it is strategic in a business or product sense. This has nothing necessarily to do with the strategies employed within design itself.

Design as a tool in business strategy has been commented upon by various authors. Such references usually relate to the employment of decoration upon products, or to the development of aesthetic appeal. It is now widely recognized that the kind of design work needed changes during the life-cycle of a product. This is readily illustrated by Table 1.

What has not been recognized is that there are alternative strategies possible within the activity of design itself. Thus, in the design of a novel product (as one extreme), it is possible and usual to go about the work rather differently from the case in which the product is being re-designed to achieve lower cost. Those familiar with design work will be familiar with the procedure − or strategy − of *value analysis* as used in cost-reducing re-design.

Some theoretical problems

At present there is a strong tendency to concentrate upon the method of designing as the main activity. Further, some theoreticians maintain that there is a single design strategy − "the strategy".

This is tantamount to saying that there is only one way to achieve

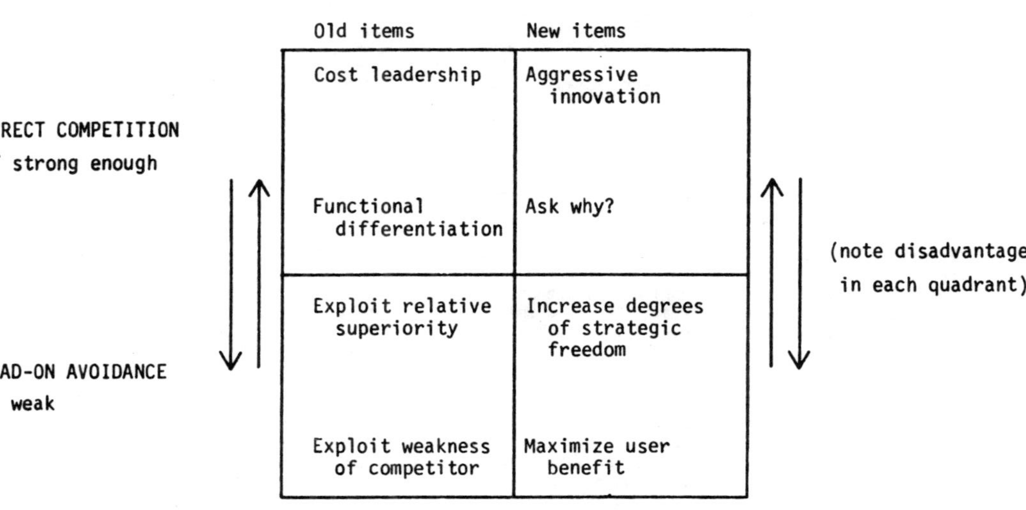

Figure 5   Competitor-directed strategy pointers

Table 1   DESIGN AND ALLIED RESPONSES WITHIN PRODUCT LIFECYCLE

	Introductory phase	Upswing (growth) phase	Levelling (maturation) phase	Post-maturity phase
Concept generation and R&D support	Radical concepts. Many attempts, prototypes. Alternative uses of available science. Special investigations. Key patents. Concern for secrecy	Intensive applied research for design and production back-up. Search for new produce uses. Extension of related basic science both for product and manufacturing process.	Continuing emphasis on R&D with emphasis on cost saving. Increase of concern with behavioural aspects. Concern for R&D control, methods used.	(decay or transformation) Withdrawal of R&D but watchful eye on concept features likely to help prolong or transform product life.
Product design	Primary emphasis on technical/function design. Imaginative leaps and rapid change. Competing design philo- sophies and strategies. No standards, some disasters.	Primary emphasis on manu- facturing process, but continuing new develop- ments in product. Pressure for standardization but with flexibility to outdo emerging competitors.	Technical change still rapid but increasing emphasis on human design. Stress on cost of design and its effective- ness. Stress on use of standards.	Tendency to 'model' type changes and minor improvements. Stress on user appeal and on inex- pensive ways of beating competitors. Some evi- dence of user re-design.
Manufacturing process design	Primarily linked to one-off or small-batch operation with highly skilled labour. Close links with product design and related R&D. Negligible economies of scale.	Move to larger batches and flow processes, if applicable. Economies of scale and rapid and reliable delivery become important.	Application of computers in process. Economies of scale affect labour needs and capital investment demands. Attention to group size and works location.	Level of output affected trade cycles and other influences. Pressures to shut down plants, using only the most efficient.
Marketing design	Reliance upon internal use, close customer linkage, attractive functional performance.	Emphasis upon increased availability. Search for profitable market seg- ments with suitable designed additions/ subtractions.	Emphasis upon design for minimum user cost. Build-up of product images for de- signed features of quality etc. Emphasis on design for purchaser lifestyles.	Low-price design, up-market design, or ongoing standard.
User	Exploratory user	Use penetration	Impact on society	Social impact on item
Investment/cash flow	High-risk, speculative. Minimum commitment of resources. Cash-flow negative.	Great demand for new investment. Problem of containing expenditure relative to inward cash-flow	Major inward cash-flow	Problems of maintaining profitability.

the desired object of design. Any reflection is sufficient to grasp the obvious fact that saying there is only one way to achieve design is akin to nonsense. What we need to recall is that a strategy is itself a plan: something which is designed.

Since the product is the focus of business strategy it is, of necessity, the "valued object" towards which design is directed. It is necessary, once more, to set the valued object and its conceptual precursors at the centre of design concern.

If we go about the business of design formalistically, we start off with a tenuous set of requirements in the object and gradually build up possibilities. At length the number of possibilities is likely to become very large. In order to deal with the host of possibilities, selection processes have to be carried out. Unless there are suitable criteria, such selection is difficult.

An obvious way of building up criteria is by reference to the objectives of the business and the particular strategy held in mind for the development of the product. From these it should be possible to make progress.

Design strategies
Strategies in design are broadly like strategies in any other kind of goal-directed activity. A strategy is a plan for putting resources into a position most likely to achieve the desired objectives effectively in a given situation. Since business competition is against others, there is no guarantee of success. There are various emphases in this essentially heuristic task (Table 2).

In design we have a valued object which usually begins by being only partly defined. Our task is to convert the various design elements which are at our disposal into a specification for the valued object by means of the resources available to us in the given situation (Figure 6).

In the activity of designing we carry out a large number of phases of work which include such actions as task definition, search or generation of possibilities, evaluation and choice. The number of repetitions is likely to be greater if the object is novel or if the system scope is large. For novel objects much of the work is likely to be done unconsciously, whereas, in order to cope with complexity, it may be necessary to have some computer aids.

Early conceptual models of any plausible valued object will tend to have a considerable number of requirements to be fulfilled. This usually leads to a complex multi-objective task which, if attempted directly and in detail, is likely to be impossible to carry through. Even with the help of a computer, the number of objectives which

Table 2   Strategic emphases in customer/competitor/company approaches

CONSIDER RELEVANT DEVELOPMENTAL PATHWAYS

Customer Emphasis	Competitor Emphasis	Company Emphasis

exploit needs/wishes
decrease price
improve benefit/price
introduce new
       benefits
offer tailored
       benefits
provide new
       possibilities
offer unique
       advantages
take overall
       perspective
	exploit his weaknesses
	use relative strengths
	find key areas
	match price at least
	differentiate
	       functionally
	differentiate benefits
	offer relative
	       superiority
	increase strategic
	   degrees of freedom
	undertake aggressive
	       initiatives:
	  worldwide cost
	       leadership
	  new product
	      introduction
		exploit strengths:
		   market/produc-
		   tion/technical
		overcome/offset
		       weakness
		make profit
		exploit preferred
		      lines of
		      development
		find allies, if
		       needed

Figure 6   Valued object derivation from design elements

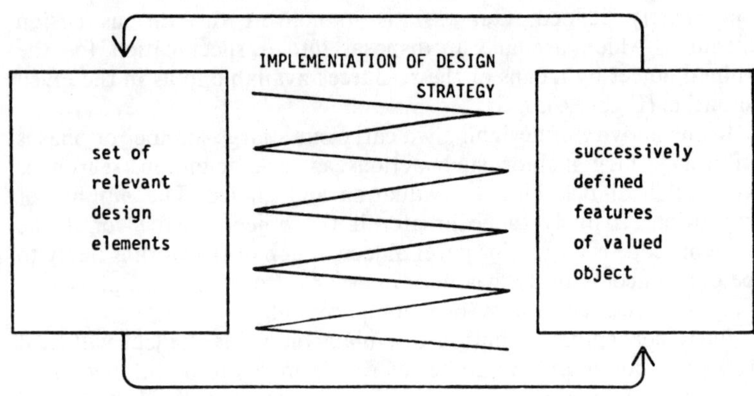

can be handled in parallel will remain small.

Design strategies tend to rely upon initial simplifications with respect to objectives as well as with respect to other aspects of design such as information. For a novel product a technical solution may initially be sufficient, although subsequently treated for cost and safety requirements. Where solutions in principle already exist, new variants may be developed by morphological analysis accompanied by low-cost product heuristics. Where social demands require safety, there are specific ways to provide safe design. Where there is a great deal of competition, particularly on the global scale, emphasis has to be put upon purchaser decision features, as in the case of high-involvement decision consumer products such as automobiles.

The initial simplification tends always to be towards the most desired objective or set of objectives. This provides provisional solutions which subsequently have to be re-worked to take account of more objectives. This usually involves the activity of trade-off.

This initial simplification is often worked through by experience, or by identifying the selection criterion in advance and only feeding into the search or generation process suggestions which, of their nature, satisfy the criteria.

The span of objectives
Although simplification of objectives is counselled in an initial approach to design, it is necessary never to forget the span of objectives which have to be achieved by the valued object. It is convenient to see them in terms of customer/competitor/company categories or something similar.

In Figure 7 a diagram is given which has been constructed to direct attention to the various perspectives which have to be comprehended. Behind everything is the set of goals which have been developed or have emerged with respect to the originating company itself. To achieve these goals an adequate pathway has to be mapped out or indicated. This is termed the "enterprise pathway" in the diagram. To bring this pathway into practical effect, a satisfactory business strategy has to be formulated and implemented.

The product itself has to be manufactured, distributed, marketed, and delivered. Each of these activities has to be carried through in ways which are in accordance with the overall business requirements as far as possible. The manufacturing process itself has to be designed. The product, in order to be acceptable for its purpose, has to fulfil certain minimum functional requirements.

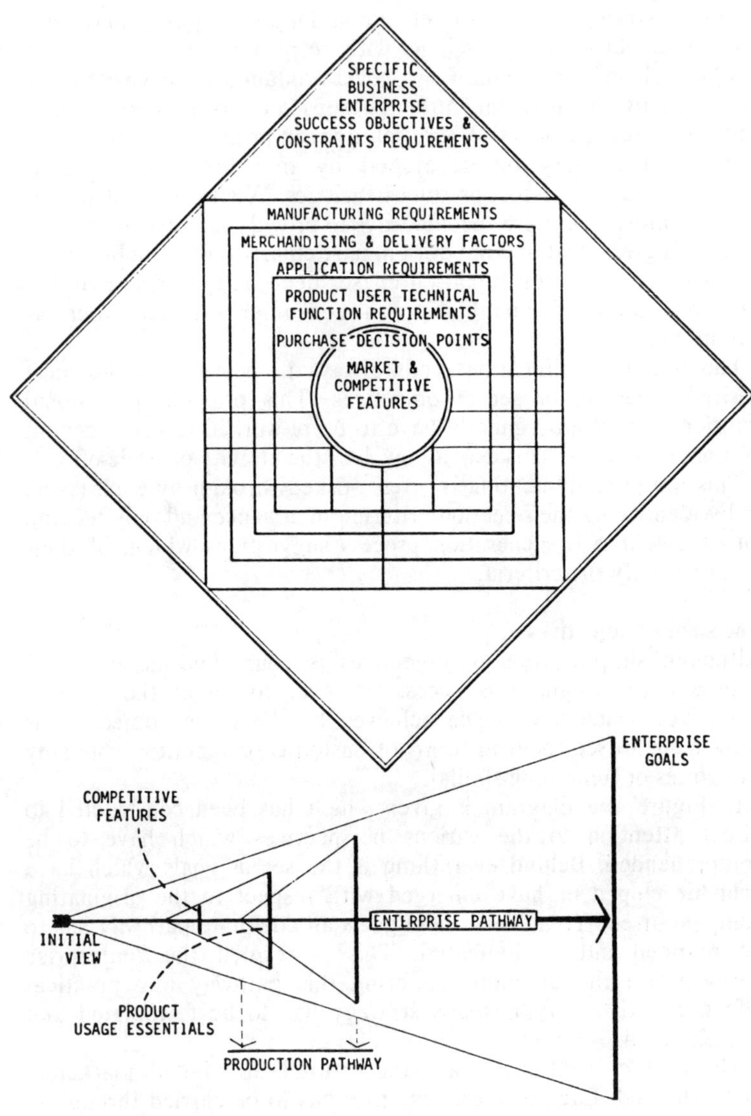

Figure 7   GOALS & PATHWAYS

Often these functional requirements are specified by national standards or by primary purchaser standards if a major distributor is involved.

Right at the front for design considerations are the market and competitive features. The design of a product has to begin with these at the centre of the scene. Unless there is a potential market there is little point in commencing design. If there is a potential market, then the product has to be able to compete effectively therein.

Entry into a market
If the initiating company does not already hold a market position, then entry can be made by way of:
- a product involving a novel functionality
- a product new to that market (e.g. already marketed elsewhere)
- making a specific extension to the market by a differentiated product
- finding new uses for an existing product design
- a product aimed to take a market share from competitors already in the market
- linkage with a company already in the market
- takeover of a company already in the market

Many examples can be brought to exemplify the various modes of entry. The automobile and computer industries are rich sources of such examples.

Some strategies in design
Gregory has recently surveyed some of the more general approaches employed in design for the purpose of identifying the locus of origin and the channels of transfer of such design technology. Among these are important strategies or components of strategies for cost reduction and novelty production. In addition there is also one strategy for the effective conduct of design work (the systems approach) which is neither aimed at low cost nor at novelty.

It could be argued that most of the set of strategies surveyed, although ostensibly dealing with design, are general problem-solving strategies and therefore almost obliged to fit into generic business strategies (Table 3).

This is tantamount to saying that if there is an identifiable business strategy, then there must be (or there is likely to be) a corresponding design strategy which is strongly related.

If we look afresh at Table 1, there appear to be some gaps in our battery of design strategies when compared with Table 3. A first

Table 3   List of well identified general approaches

Brainstorming	Operational research
Costing for design	Overall domain-specific design procedures
Design heuristics	Reliability studies
Functional analysis	Synectics
Hazard analysis	Systems engineering
Morphological analysis	Value analysis

comparison suggests that design for product differentiation is inadequately treated. To pursue this theme it is necessary to explore further various aspects of product differentiation.

Some of the ways in which product differentiation is pursued in the automobile and computer industries are set out in Table 4. For these there are design strategies.

In reviewing Table 2, it becomes obvious that design for supplier profitability also requires further consideration. At least ten strategies may be identified which deal with this area, ranging from design for low investment cost to the supplier to design for optimal profitability.

If we assemble the various approaches to design which have been proposed and discussed, there are something approaching 40 different schemes which may be listed. Of these the major proportion is that which deals with product differentiation. As may be seen from Table 4, most of these deal with what may be termed "value creation", i.e. the addition of value to products or services which already exist in principle but which require re-design in order to achieve the extra value. To some extent this suggests analogies with value analysis, but with critical differences.

Value creation has not been a subject of specialist attention up to the present, although inevitably and as a matter of course it has been practised. There is a need to put it on the same kind of basis as prevails for novelty production or cost reduction.

Conclusions

At the level of the company, practical design has advanced to some extent beyond the more or less customary level of academic preparation for technical design within a specific domain. There has been clear interest and apparently widespread usage of general strategies directed at:

- novel design
- low-cost design

Up to the present, however, there has been little obvious appreciation or discussion of the possible relationships between business strategies and design strategies. Interest in business strategies has switched from problems of integration, acquisition and disposal, and now focuses upon survival and competitive activity.

It is at this point that design strategies, originally conceived in terms of achieving adequate technical design solutions, become more directly related to business strategies by way of interest in novel or low-cost designs. This has been a shared preoccupation in

Table 4   Some approaches to product differentiation

1 Cost reduction with lower technical performance but special capability

        personal use
        compactness
        portability/pocket-sizing etc.
        add-on to existing systems

2 Quality at a given cost level

        appearance
        finish
        general handling
        user friendliness
        reliability
        life-cycle cost

3 Improved value at a given cost level

        variety in finishes, add-ons, etc.
        customer bracket specificity
        quality of user experience
        user life-style generic tailoring

4 Improved technical performance at a given cost level

        power
        speed
        extra functionality

5 Enhanced performance and value, with price increase

        ruggedness
        extra reliability
        resource economy
        built-in response to specific user needs
        user life-style specific tailoring

Germany and has helped lead to collaborative work between industrial and academic experts in the preparation of the relevant VDI Guidelines (2222 and 2235 respectively).

In some design domains, particularly those dealing with large systems, designers have been concerned for many years with questions of supplier profitability, and this has been reflected by members of the academic community.

The analysis of competitive strategic behaviour has led to the identification of generic strategies whose use depends upon the structure of specific industries, status of individual companies, and product life-cycle phases.

Comparison of generic strategies and previously identified design responses to produce life-cycle behaviour suggests that insufficient attention has been given to design for product differentiation, an approach which is needed by many companies. Detailed exploration has revealed a significant list of alternative ways of supporting product differentiation. This list offers a challenge to further work and study. In this particular emphasis should be placed upon *value creation* as opposed to *value analysis*.

Notes

1. C. Freeman, *Design and British Economic Performance*, Department of Design Research, Royal College of Art, 1985.
2. R. Rothwell, K. Schott and P. Gardiner, *Design and the Economy: the role of design and innovation in the prosperity of industrial companies*, Design Council, 1983.
3. K.G. Corfield, *Product Design*, NEDO, 1979.
4. V. Walsh and R. Roy, *Plastics Products: good design, innovation and business success*, Design Innovation Group, Open University, 1983.
5. C. Flurscheim, *Engineering Design Interfaces*, Design Council, 1977.
6. F. de Woot, *Le management des groupes industrielles*, Economica, Paris, 1984.
7. W.R. Ashby, *An Introduction to Cybernetics*, Methuen, 1956.
8. P.R. Lawrence and J.W. Lorsch, *Organization and Environment*, Harvard University Press, 1967.
9. J.R. Bessant, PhD thesis, University of Aston in Birmingham, 1978.
10. G. Johnson and K. Scholes, *Exploring Corporate Strategy*, Prentice Hall, 1984.
11. M.E. Porter, *Competitive Strategy*, The Free Press, 1980.

COMPANY STRATEGIES FOR DESIGNING AND MARKETING NEW PRODUCTS IN THE ELECTROTECHNICAL INDUSTRY

Harry Nyström

In this paper company strategies for designing and marketing new products in the electrotechnical industry will be discussed. The basis for the presentation is a framework for describing and evaluating product development strategies which has been developed over a ten-year period. So far empirical studies of more than one hundred companies in various industries (industrial electronics, industrial chemicals, pharmaceutical drugs, farm machines, wood and paper, steel, and food) have been carried out in this research programme (Nyström, 1977; Nyström and Edvardsson, 1981; Nyström and Edvardsson, 1982; Nyström, 1984).

In spite of the wide range of companies and industries studied, a number of general conclusions can be drawn with regard to the determinants of market, technological and commercial success for new products. Regardless of research intensity and technology intensity, success in designing and marketing new products seems to be highly correlated with more open, flexible and adaptive technological and marketing strategies. When the research programme was started in 1972, no framework or methodology could be found in the literature for describing and analysing product development strategy. A major concern therefore was to develop and empirically test such a framework. In this paper company strategies for finding and designing electrotechnical products will be discussed against the background of a detailed case study of a very young and successful Swedish company, Inter Innovation. In this context the overall framework for analysing product development strategies will be utilized and comparisons will be carried out both with other electrotechnical companies in the data and the overall results for all companies.

General framework

Strategies for finding and designing new products are divided into three sub-strategies: technological, marketing and organizational. Intervening between these strategies and product design are company structure and climate and organizational and psychological processes which together influence the success of product development.

A distinction is made between intended strategy, that is policy, and realized strategy, that is the pattern of regularities in actual behaviour over time (Nyström, 1970; Mintzberg, 1976). While both are analysed in the framework, the main focus is on realized strategies, and intended strategy is mainly used as an indication of to what extent companies are aware of and explicitly concerned with guiding strategic activities. Another important element in the employed definition is that strategy is mainly seen as creating favourable conditions for success in product development, rather than directly determining what specific technologies, markets and products should be developed. In other words, strategies are viewed more as instruments for promoting decentralized creative action than as centralized plans for desirable outcomes, which is the more common approach in the strategic planning literature.

Two main strategic dimensions related to success in technically designing new products, that is main components of technological strategy, have emerged from our empirical studies. The first is the R&D orientation of a company and the second is technology use. Companies which to a large extent rely on outside cooperation, e.g. with other companies, universities, research institutes or consultants, are said to have a more external research orientation, and companies which rely more on their internal competence and resources for finding and designing new products are said to have a more internal research orientation.

Our data shows that, regardless of industry, companies with more open cooperation with the outside research environment, that is more external R&D strategies, almost always have been more technically successful in designing new products. Network positioning, gaining access to external information and assistance by having wide and flexible contacts with the external environment, is therefore one of the most important strategic variables in our analysis. The types of contacts and cooperation vary between industries and companies, but most successful companies have realized the importance of broadening their knowledge and competence base by being open to the outside environment.

The second main strategic dimension in technological design

strategy is technology use. A technology is viewed as a relatively well defined and delimited area of knowledge, usually the basis for professional and educational specialization. On a fairly high level of analysis, electronics may be viewed as a technology, while on a lower level microelectronics may be singled out for attention. What level to employ in a specific analysis has to be determined in an empirical context, and the procedure in our research has been to let the informed users, the companies themselves, define what technologies they are working in.

From a technology strategy point of view companies may then stress working within given technologies, for instance electronics, or in the intersection between different technologies (e.g. optics and electronics). Both the history of technology and our research shows that major new products are usually based on what in our terminology is called synergistic technology use, the combining of previously unrelated elements of knowledge from different technologies. Our data shows that in almost all industries and companies synergistic technology use tends to be associated with a higher level of technological innovation for new products than isolated technology use.

In our framework, technological design is marketing and marketing is technological design. To successfully achieve this integration on the company level, companies need to use flexible and focused organizational mechanisms. Our data shows that successful companies almost always use a project organization in product development to achieve this goal. In product teams, flexibility and diversity is usually brought about by a mixture of team members with different backgrounds, e.g. from both marketing and R&D, and a change in membership over time, e.g. with more R&D specialists at an early stage and more marketing people at a later stage. Integration and direction is generally sought by strong professional project leadership and detailed financial budgets and time- and performance-related objectives. In small, highly innovative companies, the whole company may be run in project form with the manager as project leader, and R&D as the major function.

From the point of view of our analysis, the critical role of recruitment strategies for long-run success in product development is also apparent. Particularly in small and highly innovative companies working in new technology areas, the most crucial issue in developing technological competence is often finding competent personnel to strengthen and broaden the knowledge base. This largely determines both the need for and possibility of employing different strategies for technology use, and recruitment strategy is therefore an important aspect of product development in our approach.

Last, but not least, success in designing and marketing new products depends on the organizational climate prevailing in a company. A favourable climate based on meaningful challenge, intellectual freedom, mutual support and tolerance for differences in thought and action, are found to characterize creative companies in our research (Ekvall, Nyström and Waldenström, 1983). Due to the complexity of climate we have not been able to relate changes in company strategy directly to changes in climate. Instead we employ measures of existing climate as an intervening factor which in our framework influences the effect of our strategy variables on product development.

A case study of successful innovation

Inter Innovation is one of the most successful innovation companies in Sweden. It was started in 1973 by Leif Lundblad as a one-man inventor company and in 1982/83 had 450 employees and an annual turnover of 150 million Swedish Kronor. The founder had previously built up a successful business in renting construction machines and sky lifts and cleaning buildings. Based on his technical interest and early experience in running a television service shop at the age of 17, Leif Lundblad wanted to invest his profits from his earlier ventures in an innovation company. His initial objective for the company was buying and selling licences and acting as an intermediary between independent inventors and companies. This approach was not successful and instead he started to develop and commercialize his own ideas.

The company's first product was a money cassette for use in petrol stations in connection with automatic payment systems. This made money handling and control much easier and the product was successful. At the same time self-service cash dispensers were being developed for banks and Inter Innovation realized that they could develop their basic technical design from the money cassette to solve the cash dispenser needs of automatic bank service. Inter Innovation tried to commercialize this idea by licence agreements with an English and a German company. This cooperation was not successful due to insufficient interest of the companies in promoting and commercializing the prototypes provided by Inter Innovation.

Instead, Inter Innovation decided to develop and market its own product, a multi-denominational cash dispenser. This was done in close cooperation with the first customer, Citibank, in the USA.

The result was an order for 20 machines and a letter of intent for 761 machines of an as yet undeveloped product, with very strict technical buyer requirements. Inter Innovation now needed to finance and technically produce almost 800 "Cash Adapters", as the new product was named, with no production facilities of their own and a capital base in the company of less than one-tenth of the price of a single machine.

Again Inter Innovation chose to cooperate with established companies in producing and selling the products, but had now learned to retain complete product control. All offers to buy licences from Inter Innovation, or the company itself, were resisted, and instead Inter Innovation acquired a manufacturing company and very rapidly converted it to produce its own products.

Inter Innovation is now established as a successful but still highly innovative company with subsidiaries in many countries such as England, Germany, Spain, France and the USA. It devotes 10% of sales to product development and is diversifying into new technology-related product areas, such as money handling in retail stores.

What then are the secrets which explain the oustanding success of this company? The overall reason is the realization that product design is marketing and marketing is product design. The company has been run basically as a marketing project, with product development strategy as the leading line of development. This strategy, in turn, may be characterized by three main attributes: speed, flexibility and cooperation.

From the interviews and written documents which the case study is based on, it is clear that Leif Lundblad has emphasized the three strategic elements noted above, in the very dominant leadership role he has assumed. In his own words, the main lesson he has learned is the need to cooperate with buyers and other companies in product development and design, without giving up product control.

Licensing is ruled out as an active cooperation mechanism for constructive product development, since the company then loses control of the development and marketing process. By letting buyers help to design and test products and pay for prototypes, cooperation and buyer commitment is achieved, without any loss in product control for Inter Innovation. By maintaining a flexible organization, the company then can quicly adapt to changing conditions and stay ahead of competitors.

Until recently, Inter Innovation was basically a project organization, with Leif Lundblad functioning both as Managing Director and Project Leader. This led to a close integration of different functions and high responsiveness to buyer needs, with Lundblad dealing directly with all aspects of R&D and marketing, both within the company and in relation to customers. With the rapid growth of the company this type of management has become increasingly difficult, but it no doubt has been an important factor behind the company's success.

Equally important has been the emphasis on speed in product development and commercialization, which of course has been made possible by the driving initiative of the leader and his insistence on organizational flexibility. Carrying out parallel activities and starting new projects while phasing out old ones, together with constant experimentation and a willingness to reassess quickly and learn from mistakes, has been the dominant leadership style. This has led to an intense and dynamic atmosphere in the company and an exciting organizational climate which has attracted creative people but at the same time proved unacceptable to those used to working under more orderly conditions.

By setting seemingly impossible time limits and goals, Leif Lundblad has succeeded in speeding up development time, which is one of the most crucial competitive parameters in highly dynamic industries such as electrotechnology. Selling products before they are developed, by showing prototypes, is one way of gaining momentum. Another way also emphasized by Inter Innovation is stressing flexibility in product design by initially designing and producing multipurpose modules, which at a later stage of the development and production process may be assembled to serve specific buyer needs. This facilitates both the adaptation of existing products to specific buyer needs and the development of new products, while providing better economics in production and distribution.

A further aspect which is evident in Inter Innovation's product development strategy is the avoidance of technological sophistication as a goal in itself. The company has not fallen into the technological trap of isolated technology use, namely searching for the most advanced solutions within given areas of technological specialization. Instead, they have looked more to synergistic technology use, by combining conventional electromechanical and computer-based knowledge to achieve simple, cost effective and functional design. This has been an intentional strategy promoted by Leif Lundblad and influenced no doubt by his own lack of formal technical training, and independent, practical approach to product development.

General implications for strategy

An attempt will now be made to generalize the implications of our case study by utilizing our framework and the results from studying product development strategies in other electrotechnical companies (Nyström, 1977). To begin with, electrotechnology is a rapidly advancing and highly competitive technology. This means that it is difficult to achieve patent protection and to prevent competitors from imitating successful new products. Instead it is mainly by maintaining a technological lead that companies can achieve a competitive advantage. Competence in designing new products and quickly marketing them, even before they are fully developed, becomes a crucial element of success.

This also means that external R&D orientation, for instance joint product development with customers, becomes a highly attractive element of strategy (von Hippel, 1978). Buyers then become committed to a product at an early stage of its development by influencing the process. If joint development can be combined with advance orders or other contractual sales arrangements, such as letters of intent, buyer commitment of course is further enhanced, and also the likelihood of commercial success in product development.

The rapidly developing state of the art in electrotechnology and related areas of knowledge means that an external R&D orientation towards the scientific community at large, universities, research institutes, consultants and inventors, becomes highly desirable to a company if it wants to maintain a strong technological position. Cooperation between different specialists within a company also becomes important to promote synergistic technology use, that is the combining of knowledge from different areas of technology. Combining electrotechnology with computer programming is an example of this, where the combination is more important than excellence in either area. In Inter Innovation such coordination is achieved by direct intervention from the managing director, also acting as head of R&D. In another electrotechnical company studied, more decentralized coordination was facilitated by making different types of specialists work together at the same desk, and by frequently moving people around in the laboratory.

The necessary integration of technological and marketing elements in electrotechnical product development is also facilitated by flexibility in product design. This is usually achieved by standardizing components and achieving product differentiation and adaptation to individual buyer needs by combining systems of standard components. Selling general design, rather than specific function, by using simple cardboard prototypes, is another aspect of flexibility in product design successfully employed by Inter Innovation. Another strategy used by electrotechnical companies is to make contractual commitments to deliver products which fulfil specific functional requirements, without evenhaving prototypes available. This reduces direct competition, but introduces great technological uncertainty instead and strong pressures to meet time and cost criteria. This means that the technological competence and image of the company becomes more important as a competitive parameter than specific product design. In other words, marketing then becomes based more on knowledge and competence and less product specific.

Our strategy dimensions consider the need for building up the company's overall technological competence and potential and not only product-specific strengths and weaknesses which are usually the focus of analysis in traditional marketing approaches to product development (Pessimier, 1977; Urban and Hauser, 1980).

At the same time of course, technological strategy and competence to a large extent are influenced by and influence the company's overall marketing strategy. A more concentrated marketing strategy means fewer and more similar products and customers, while a more diversified marketing strategy implies more numerous and dissimilar products and customers. A more open technological strategy, based on more synergistic technology use and external R&D orientation, should usually be more necessary and successful when pursuing a more diversified marketing strategy. Highly successful young technology-based companies such as Inter Innovation usually have a fairly narrow technological base and a highly concentrated marketing strategy. Because they usually have unique products in new market areas with high sales potential, they do not typically experience a very high need for a more diversified marketing strategy. Instead they may, like Inter Innovation, employ synergistic technology use and external orientation to achieve a more successful, concentrated marketing strategy.

Larger, more established companies, on the other hand, may find it desirable to use synergistic technology or external orientation to achieve a more diversified marketing strategy. An example of this among large Swedish companies in the electrotechnology industry is ASEA. They have successfully used the combining of technologies to diversify into industrial robots and joint technical product development, and together with the Swedish National Railway to diversify into electrical locomotives.

For competitive reasons, speed and flexibility in designing and marketing new products thus are of vital importance to electrotechnical companies, more so than in most other industries. Even in such a high-research and technology-intensive industry as pharmaceutical drugs, companies are less dependent and able to achieve rapid product development because of the greater opportunities for product protection and legal restrictions on introducing new drugs. Because of the great market instability and high rate of technical change in the electrotechnical industry, companies must constantly try to be technically ahead of their competitors, but also in phase with their customers. This is best achieved by network positioning, systematically managing relations with customers, other companies, universities, consultants and other sources of environmental support. A variety of strategic measures for achieving this type of flexible, interdependent development have been discussed in this paper. Basically these strategies are aimed at promoting creativity by widening, combining and focusing knowledge in flexible ways to achieve constructive results.

References

Ekvall,G., Nyström, H., Waldenström, I. (1983)
   Organisationsklimat och innovativ förmåga, report från FA-rådet, Stockholm.
von Hippel, E. (1978)
   "Successful new products from customer ideas", *Journal of Marketing*, no.42 (January), pp.39-49.
Mintzberg, H. (1976)
   Patterns in Strategy Formulation, report from McGill University, Montreal.
Nyström, H. (1970)
   Retail Pricing. An Integrated Economic and Psychological Approach, Norstedts, Stockholm.
Nyström, H. (1977)
   Company Strategies for Research and Development, report from the Department of Economics and Statistics, SLU, Uppsala, Sweden.
Nyström, H. (1979)
   Creativity and Innovation, Wiley, New York/London.
Nyström, H. (1984)
   Product Development Strategies in Swedish Paper Companies, report from the Department of Economics and Statistics, SLU, Uppsala, Sweden.
Nyström, H. and Edvardsson, B. (1981)
   The Importance of R&D Cooperation Strategies for Product Development, report, Dept of Economics and Statistics, SLU, Uppsala, Sweden.
Nyström, H. and Edvardsson, B. (1982)
   "Product Innovation in Food Processing: a Swedish survey", *R and D Management*, no.2 (April), pp.67-72.
Pessimier, E.A. (1977)
   Product Management - Strategy and Organization, Wiley, New York/London.
Urban, G.L. and Hauser, J.R. (1980)
   Design and Marketing of New Products, Englewood Cliffs, New Jersey.

AUTOMOBILE USAGE
IN A FUTURE INFORMATION SOCIETY

Ove Svidén

This article is based on a Swedish project study which is part of the Future of the Automobile Program initiated at the Massachusetts Institute of Technology in 1980.[1] Entitled "Automobile Usage in a Future Information Society", the Swedish study was performed at the Institute of Technology, Department of Management and Economics at the University of Linköping. The study was financed by a number of government agencies concerned with road and traffic administration and technical development.

The purpose of the study, and of this article, is to investigate the effects of improved information technologies on automobile technology and usage, with Sweden being chosen as the main example. Scenario scenes for the years 1990, 2000, 2010 and 2040 indicate some of the structural changes necessary for an industrial society to evolve into a mature information society. The scenario is used as a base for a quantitative estimate of travel demand and automobile usage in Sweden in the future, with 1980 as a reference year. Some of the results of this futures study are intended as input to the Swedish government's long-range planning.

Cars and the future

The future post-industrial society is an information society and its evolution implies major structural changes.[2] It will influence the way we live, work, study and use our leisure time. Automobile usage will also change. On the one hand automobile transport can, in some cases, be replaced: parts of our work can be performed from our homes or from offices within walking distance of our houses, using high-quality tele-communications, decreasing the amount of daily commuting; some of our shopping can be ordered via telecommunication channels and delivered by vans to our homes. On the other hand, improved information services and networks can also result in new business contacts and vacation demands that can increase travel and automobile usage.

The combination of a high-quality information network and good transportation can make possible a more dispersed living pattern with decentralized industrial organizations and governmental functions. In a society a high information-mobility may become a counterforce against urbanization. The information society can be a society with small towns and rural living, linking its inhabitants by high-quality telecommunication with local, regional, national and international activities.

Impact of new technologies

A more direct influence of the development in information technologies (electronics and microcomputers) can be seen in automobile performance and road information systems. Present improvements regarding engine emissions, fuel efficiency, power conditioning and road navigation often rely on micro-electronics developed during the last ten years. Modern telecommunication technologies and speed control functions can be used to improve road capacities and make the traffic flow smoother and safer.

If automobiles in the future information society have a "programmed traffic behaviour" the demand on user driving skill will be reduced. The automobile could become a transportation means for new groups of users, for example teenagers, elderly and handicapped people. The future ecological automobiles will have to be equipped for safer and more efficient traffic behaviour. It will become natural to most automobile users not to own an automobile but to lease or rent it. Automobile fleet-owning organizations will have a very important function acting as "competent customers" of the automobile industry. They will have the power to specify the technical standard of the automobiles and their life-cycle costs.

In the information society road planning will have to be tied to the development of the vehicles. This means that the road transportation system, including vehicles, roads, data links and computer functions in automobiles as well as in roads and in traffic control centres, must be seen in a system context from planning and development to its step-by-step implementation.

There is also a conflict between automobile usage and living in cities, as Gabriel Bouladon has clearly stated:[3]

The present day city mirrors the internal conflict of modern man. He wants to live in towns and preferably in towns which are pleasant to live in. But at the same time he does not wish to be deprived of his car, and indeed cannot do without it. Yet the present day automobile has shown itself to be incompatible with the city. Is there any hope of resolving this dilemma in the next generation without endangering a powerful automobile industry, probably vital to the survival of our industrial society ...?

How can this conflict between dwelling and mobility be solved? By better information and communication technology that will make unnecessary some of today's automobile travel? Or by the development of the automobile into an ecological vehicle, clean, silent and "inherently safe at any controlled speed"? An ecological car is "rigorously non-polluting", has a long life and is "almost maintenance free and capable of being driven by anybody", according to Bouladon. Can the automobile industry produce this ecological vehicle? Or, finally, can the conflict be solved by a synergy between higher information standards *and* higher mobility? Is this synergy a counterforce against cities? Can it, for example, be used as a policy or a tool for an evolving decentralization? What capacity increase and which safety improvements can be the result of semi-automatic highways?

Scenario method
For the futures research in this study a scenario method has been selected. As Calder has written: "The aim is not to prophesy about what the actual future will be but rather to rehearse futures accessible to political choice."[4] According to the Oxford English Dictionary a scenario is "a sketch or outline of the plot of a play, giving particulars of the scenes, situations etc. ... Scenarios usually consist of a verbal description of a hypothetical situation at a future point in time" and they can include a sketch of the main changes which are assumed to have taken place in the intervening period.

The scenario method used for this study differs somewhat from the type of low-growth/high-growth quantitative scenarios frequently used. This scenario is a qualitative description of society at different points in time. It represents a synthesis of many different ideas, and the scenes indicate the time needed to restructure the present society into a mature information society. Sweden is chosen as an example. The scenario scenes selected represent:

- A "crisis" situation during the latter part of the 1980s, limiting energy supply to transportation.
- 2000: a society with a high economic growth and a vital restructuring of industry. New ecological and informational technologies are developed.
- 2010: an information society in transition, one generation ahead.

- 2040: a mature information society with decentralized organization and dispersed living, two generations ahead.

The scenario scenes represent a base for the quantitative estimate of travel demand and automobile usage which is presented later.

"Crisis" scenario
The economic indicators in the early 1980s are not encouraging. The recession of 1982 continues and deepens into a depression in the following years. A number of military conflicts around the world in the late 1980s result in disruptions of oil supply to OECD countries.

The slowdown of consumer industry is partially compensated for by a number of large government crisis programmes. The overall picture is that of industrial production reduced to 60-80% of normal output, leading to a reduction of the working week by one or two days in many industries. The slowdown of industry and the efforts at energy conservation by reducing travel lead to a situation where working part-time at home has become a virtue. It saves fuel for transport and for heating. With a reduced working week the established routine production is maintained rather efficiently. At the same time a part-time working force of professionals is available to act as consultants for the government crisis programmes.

Automobile usage for free-time travel is restricted and reduced drastically. Commuting, business and service travel have to be done in a more efficient way by car pooling, van pooling and by public transit. It is possible to achieve a 50% reduction of fuel for transport by these measures. A great deal of business has to be performed by means of telecommunications. Telephones and the postal services are used to their limits. Systems for car pooling are developed and tested. Home terminals of the mechanical keyboard/cathode ray type are being used extensively. The imperfections of the telecommunication systems lead to irritation and thus to creativity, which aids the development of improved information systems and services. The crisis also leads people to a number of new work experiences and insights about the need for mobility and improved information technology.

"2000" scenario
In the year 2000 we can see a healthy growth in the world economy once again. A new, positive and strong belief towards development and in new system solutions has appeared. New organizations and industrial relations evolve, a growth in new ecological technologies can be noticed and new information networks and services are being

developed at a fast rate.

Automobiles are to a large extent rented and leased, rather than owned by the users. The rental companies have achieved an important economic strength. Transport authorities have developed a systems view and much effort is spent on developing an efficient transportation system for society.

The automobile industry is becoming increasingly transnational. Meanwhile, engine development becomes separated from the automobile industry, with the engine being regarded as a component. Large-scale R&D efforts are directed towards new engine system concepts; hybrid technologies are used. The new "gasistor" engine concept looks promising; its continuous combustion gives low emissions and allows the use of a wide range of fuels.

The working situation has stabilized well after the crisis in the 1980s. There is now a goal-oriented rebuilding of industrial production. Decentralization is a key word in the rebuilding of industry, homes, energy and service infrastructures.

Professionals are in high demand for projects domestically and internationally. A large proportion of them have voluntarily chosen a working situation with more than one employer, i.e. working as consultants from their homes, and at local offices some days of the working week. It becomes increasingly more common to make this shift when reaching the age of 40; some professionals made this choice earlier, during the crisis in the 1980s, and preferred to continue this way in the following decades, thus setting the pace for others.

"2010" scenario
By the year 2010 the information technology available had been developed to such a standard that information on screens is preferred, replacing much of the information previously carried on paper. Most routine office work is performed via "intelligent" terminals. Home terminals can easily be connected to office files and thus office work is no longer restricted to specific working hours or geographic locations.

High-resolution flat screens are mass produced and are found in place of TVs, radios, telephones and typewriters. Daily news is screened through a filter programmed by the individual, giving in-depth information only on certain specified issues. The flat screens are as small as a book or magazine, and begin to replace many of these publications.

Other screens are of the large, wall type, giving a cinema

atmosphere to the home when so desired. This type of presentation is preferred by adults and elderly people with reduced vision. Browsing through an entertainment video, shopping advertisements, an article or a memo is more convenient on the wall screen than by using the small screen or browsing through bundles of paper. Tele-data terminals (TEDAs) are equipped with computer logic for word processing, spelling, calculations, filing and retrieval. Other sub-programs are used for editing, diagram production and translation. Short-distance communication is mostly done by glass fibre optical cables. About 50% of the economically active population now has access to high-density data transmission lines.

The road traffic control centres equipped with computers and "green wave programs" are in data-link contact with about 20% of the vehicles on the road. The effects on traffic are noticeable; the automobiles equipped with speed command functions act as moderators, making the traffic as a whole flow more smoothly. Information technologies improve the use of limited resources. Thus travel planning, route selection, car pooling, van pooling, electronic hitch-hiking and private taxi services can be coordinated in an efficient way. Small cars are no longer very efficient. The smaller car for shopping may be needed, but the larger car with capacity for pooling is more economic.

"2040" scenario

In the year 2040, about two generations ahead, one foresees a mature communication technology serving information needs well. TEDAs, of many models and shapes, are as common as TVs and telephones were in 1983.

More than 70% of the economically active population in the information society is dealing with information of some variety or other during most of the working day. Environmental concern and the trend towards decentralization lead to a dispersed living and working pattern, homes being spread out along rural roads.

Roads themselves now define a geographical domain, a network, not only for transport but also for the distribution of energy and information. Pipes for natural gas and fresh water supply run alongside cables for electricity and information beside the roads. Almost every house is connected to these services. In more densely populated areas the houses are also connected to sewage and district heating systems.

Local information centres are in communication with the international information society via glass fibre optics and satellites. Information and mobility are the cornerstones of society in 2040.

The dispersed living and working pattern is possible due to synergy between the two.

Automobiles are semi-automatic, making mobility accessible to almost all people above 14 years of age. The information technologies and automation have reached such a high reliability that automobiles now can be regarded as "safe at any controlled speed". Many elderly and handicapped see their mobility by automobile as one of the blessings of their time.

While "information" work is performed mostly at home and in local offices, much of the production and manufacturing is done in many small, decentralized, automated factory units. Local manpower is employed to operate these units, and to monitor and service the manufacturing robots. Work is a part-time duty, a necessary and boring task, that can be accepted as long as it does not interfere with people's part-time education, research and development work. Creative and artistic work is held in high esteem.

Automobile transport is a necessity. Dispersed living and working patterns lead to a situation where public transport duties are requested from automobile drivers. Advanced forms of rapid pooling of transport demands and transportation resources (such as automobiles and vans) are available through the TEDA terminals. A businessman on his way to a meeting may agree to pick up a paying passenger if it does not mean a detour or delay of more than two minutes. A shopping mother in no immediate hurry may agree to give a lift to a handicapped elderly person, resulting in a detour of ten minutes, as she is adequately compensated for the service performed with an automatic money transfer from the social security. A student or an elderly person may spend a couple of hours a day as a taxi driver. These types of para-transit services performed by automobile drivers may in the future completely replace public transport services in "trans-urbia", i.e. outside larger cities. Thus the automobile in the information society will be used in a multitude of passenger transport roles, administered effectively by the information technologies available.

Travel demand in the future information society
Using these scenario scenes as qualitative bases, the total travel demands for each period have been estimated with Sweden as the example. The results are shown in Table 1.

The base line is the travel situation in Sweden in 1980, which is based on a major traffic measuring programme undertaken in 1978. This revealed that about 80% of all travel was performed by

TABLE 1. A QUANTITATIVE ESTIMATE OF FUTURE TRAVEL STRUCTURE AND AUTOMOBILE USE IN SWEDEN.

S t r u c t u r e G r o w t h		Scenario scenes				
	Travel type (%)	**1980**	**'Crisis'**	**'2000'**	**'2010'**	**'2040'**
	Commuting	21	15	17	19	17
	Business	15	10	13	15	15
	Service	11	10	10	6	6
	Free time	45	10	40	45	30
	Miscellaneous	8	5	5	10	12
	Travel distances, relative to 1980	100%	50%	85%	95%	80%
	Relative population index	100	100	110	120	144
	Relative travel distance index	100	100	110	132	175
	Relative passenger kms	100	50	102	150	201
	Auto passenger kms/year: 10^9	65	32	67	97	131
	Automobile load factor					
	(average number of people/car)	1.3	2.0	1.6	1.4	1.5
	Automobile kms/year: 10^9	50	16	42	70	90
	Total automobile energy					
	consumption relative to 1980	100	30	80	125	160
	Number of automobiles in use					
	relative to 1980	100	70	100	130	160

automobile. Air, train and bus traffic only represented about 20%. Another striking fact was that commuting only represented 21% of total passenger kilometres travelled, while free-time travel represented as much as 45%.

For the four scenario scenes, first the structural changes in travel for commuting business, service, leisure-time and miscellaneous trips are estimated for the different periods. Second, the relative increase in travel distances per person owing to more dispersed living is estimated. Third, the relative population growth at each period is considered. The total of estimated travel distances can then be calculated. The travel distances in Table 1 are presented as percentages of the 1980 situation.

Commuting
It is estimated that commuting to work can be reduced by about 20% by the year 2010 as compared to the 1980 situation. After two generations, i.e. in about 60 years, it can be assumed that a mature information society may result in dispersed living and decentralized working functions at local offices and workshops. When these living and working habits have restructured society, commuting to work may be reduced by about 40%.

The economically active people in Sweden represent 43% of the total population, or 67% of the population between 16 and 67. Also the number of people having a paid job in addition to work at home is increasing; between 1970 and 1975 this increase was 3.6%. In an information society there will be more opportunities for work. When geographical distance is no longer a barrier to work participation as it is today, a much larger part of the population can contribute. Matching individual ambitions and experience with job opportunities will be one of the major assets of the information society. This can lead to a situation where 90% rather than 67% of the population aged 16-67 are employed full time or part time.

A consequence of this will be that the relative amount of travel for work will increase about 20%. Thus the increase in travel to work due to a higher employment ratio will reduce the substitution calculated previously (20% and 40%) to, say, 10% and 20% relative travel reduction in the scenario years 2010 and 2040. It should also be remembered that these reductions refer only to commuting, which in 1978 constituted only 21% of all distance travelled.

Business travel
Business travel represented 15% of the total amount of passenger kilometres travelled in Sweden in 1980. In the future information

society, some business contacts which are today made by meeting and travel can be achieved by tele-conferencing. For seminar-type meetings where more than ten people listen to a lecturer or specialist using OH pictures, the TEDA technology may offer a substitute. Without attending the meeting physically, one can hear the presentation and have the pictures shown on a terminal. The recording function will allow the verbal presentation to be kept; the pictures can be recorded for future use as support material for a report or presentation, and material and comments can be presented to a meeting via cable, after being cleared to do so by the "speakers list" sub-program of the TEDA conference service.

However, if the information society leads to decentralization and the spread of industrial production, an increasing amount of face-to-face contacts may be needed to make the industry perform efficiently. It is probable that the substitution made possible by improved telecommunication will be balanced by an increase in face-to-face meetings between officials and workers in different geographical units in organizations. The outcome will be that business travel will be maintained at the same level in the future as it was in 1980.

Service travel
Service travel represents 11% of total passenger kilometres in Sweden. It is dominated by the purchase of goods from shops and supermarkets, but also includes visits to the doctor, community centre, nursery, library, laundry, bank and post office. The automobile is most used in service travel because it will transport door-to-door.

In the information society many service trips can be replaced by information services which aid the selection of the things to be bought, and by rational distribution by vans directly from regional stores to the home. For example, the weekly or monthly shopping of staple foods can be selected on TEDA, ordered and delivered to the door during the day, or to a local service centre in the neighbourhood, a kind of "refrigerated mailbox" service. As well as groceries, household goods, newspapers, magazines, books from the library can be ordered this way. Most bank visits will be unnecessary with the use of credit cards, home shopping and home banking.

Such methods could reduce the need for service travel by automobile by as much as 50%. But, again, people may not want to change their shopping habits just because it is possible; it may take a generation to shift habits. Therefore by 2010 the reduction in service travel may be only 25%. Also, if the information society includes decentralization and a spread of living patterns, the length of the service journeys may be increased. This will reduce some of the travel savings mentioned earlier.

Leisure travel
Leisure travel, such as visiting relatives, friends, sport facilities, evening classes, weekend trips to the country and vacation trips, account for 45% of the passenger kilometres undertaken in Sweden. The family automobile is the most important type of transport for leisure travel. Its type, size and loading capacity are usually selected to fulfil these travel demands.

Services available in the information society will probably not change much of the leisure usage of the automobile. Improved telecommunication services cannot substitute for social meetings with friends and relatives, or an evening lecture or seminar in a study group. Sport activities and weekend trips for hikes or visits to the country are equally not replaceable. The "double dwelling" pattern, with one flat or house in town and a country cottage for weekends is a popular pattern in Sweden. (The country cottage at first consisted of a simple hut for the summer holiday; it is now a well-equipped house for vacations and weekend trips from March until October — some are even permanently occupied. This double-dwelling pattern can be seen as an intermediate step between the life of the industrial society and the dispersed existence possible in the information society. The transportation demand from having two dwellings is high, and is a luxury which has to be reduced in case of energy restrictions. In the long run the demands of double-dwelling will decrease as a result of increasingly dispersed living.)

Miscellaneous travel
Miscellaneous travel is the final category. In Sweden 8% of all journeys do not fall into the categories of commuting, business, service or leisure travel.

Travel of the future
In a future information society there will be types of travel that we today lack the imagination to foresee or to find reasonable — for example, there may be evening classes in ecosophy, meditation retreats, creative politics, TEDA program production, electronic hitch-hiking. With electronic hitch-hiking a travel demand may be made through a TEDA to get a lift with somebody going in the same direction (somebody prepared to accept car pooling and a

small payment — electrically transferred, of course — to reduce his travel cost).

In an information society there are risks that people may become isolated through using a TEDA during work and free time. Some drivers may welcome the opportunity to meet new people at the same time as reducing one's own travel costs. Such instantaneous car pool arrangements, if they can be made to work rapidly and safely, represent an interesting alternative to public transit. For example, one hitch-hiker in every fourth car on the road corresponds to the used capacity of the entire Swedish public transit system!

Travel structure
The structural changes of automobile travel over the four scenario periods are also presented in Table 1. In a "crisis" situation, automobile travel could be reduced to 50% of the 1980 level, mainly by a drastic cut in leisure travel and by some switching to public transport for commuting and business travel.

By the year 2010, automobile usage is "back to normal" again, with passenger kilometres about 95% of the 1980 value. In a mature information society in the year 2040, the dispersed living patterns and decentralization of work may again result in a reduced automobile usage, due to increasing competition from efficient regional public air transport.

Population and demographic trends
The evolution of the information society includes much more than technology. Shifts in attitudes and life styles have to develop; this can be a process that can take one or two generations. Thus a projection of population growth and of demographic trends has to be included in order to quantify automobile usage changes.

It is assumed that the population will grow through immigration and fertility by about 20% every 30 years. This is about the same growth rate (18%) that has prevailed in Sweden during the 1920-1980 period. Urbanization in Sweden has been a marked trend for at least a century; but during the 1970s a break occurred as people started to settle in areas outside the suburbs and some families moved back to rural living. In other words, this recent Swedish trend favours more dispersed living and is a result of improvements both in transportation and in telecommunications, as depicted in Figure 1.

IIASA has studied the urbanization problem[5] and found that similar breaks in trends have occurred in other large city areas in highly urbanized countries. It is assumed that this new trend is not a

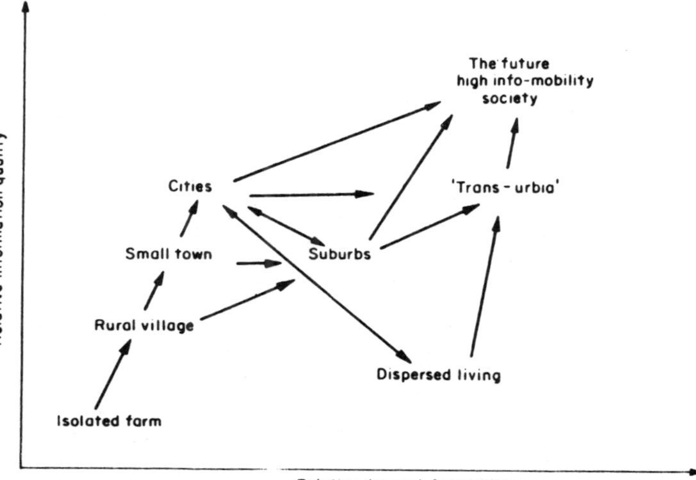

Figure 1. Future info-mobility.

Improved information services and functions can substitute some travel, but also generate more new demands for travel; improved mobility makes it possible to combine the advantages of city life and the ecological qualities of dispersed living, and both with the contact networks available today in cities only. The future high info-mobility 'trans-urbia' can be an alternative to most living and working patterns today.

temporary phenomenon, but the start of a marked shift towards dispersed living patterns. This could result in commuting for long distances; however, as the information society develops, decentralized industrial activities and government functions can also provide a dispersed pattern of job opportunities, decreasing the demand for commuting in the long run.

How much will this factor change the distance of travel? The growth of the suburbs around our cities can be regarded as a first step in the dispersal movement. Suburbanization in the past 30 years has been parallel to the development of TV, which has allowed the events and entertainments of the city to be consumed at home in the suburbs, and the rapid growth of automobile usage.

The next step towards the information society will be when home TV, our telecommunications and microcomputing technologies have developed sufficiently so that the home TV can be used as a home terminal that is as well equipped as the office terminal. Working at home will reduce daily commuting, and with reduced commuting, longer travel distances will become acceptable, and a more spread-out living pattern will be possible. The earlier suburbanization may be followed by a "trans-urbanization", with the road network beyond the city and town acting as he geographical focus for settlement.

The increase in kilometres per passenger and per day over the past 60 years gives a foundation for a forecast of the commuting decades. The increase in travel distances from 1970 to 1980 was 22%. The 1980s may be a decade of recession and depression with a moderate increase in travel distances. After 1990 the growth in travel per person is assumed to increase by 10% per decade, a slow growth compared to that of before 1970, before the stabilization of the number of cars per inhabitant. If this 10% per annum growth in travel distances continues until the year 2040, the increase then will be 75% over 1980.

Auto passenger kilometres

A multiplication of the structural factors indicated above with the projected population growth and the travel distance indices will give the figures for "relative passenger kilometres" travelled by automobile as compared to the 1980 value. They are shown in Table 1. Automobile usage is assumed to double between 1980 and 2040, so even if there is some structural reduction in automobile usage, the population growth and effects of dispersed living will dominate.

Automobile usage
The average number of people per car in Sweden today is 1.3. In a crisis situation pooling may lead to an increase in this figure to 2.0. In a mature information society, electronically controlled car pooling may result in a load factor that is higher than today, but not as high as in a crisis situation.

It has been assumed that car pooling can be undertaken successfully in a society with dispersed living. However, this would require not only suitable functions and TEDA terminals in most offices and homes, it may also need a TEDA type terminal at each bus stop! (Or perhaps the travel request will be keyed in via the wristwatch.)

Fuel consumption
Using the assumed load factors and the figures on automobile kilometres per year, the relative automobile energy consumption has been calculated and is shown in Table 1. A small improvement in fuel efficiency is included. As can be seen, the energy demand from automobiles may increase by 60% between 1980 and 2040. During a crisis, fuel consumption for automobiles may be reduced by as much as 70%, mostly by reducing leisure travel and by intensified pooling of travel demands. However, fuels for public and military transport may increase. Thus the "fuels for transport" reduction as a total may not be more than 50% in a crisis situation.

Numbers of automobiles in use
The key feature of automobile ownership and travel today is availability. It is reasonable that this will be even more so in a dispersed information society. Beside owning a car, leasing another and renting more of them when and where needed, a family in the year 2040 will see automobility as an essential part of the quality of life. Today 25% of the population are too young, and 10% too old and/or handicapped to drive an automobile. These figures may be reduced to 20% and 5% in a mature information society with semi-automatic traffic.

New markets for specialized cars like shoppers, station wagons, para-transit vans for pooling, pick-up trucks, land rovers, sport vehicles and mobile homes will develop. If automobility is a prerequisite for dispersed living, and if governments will support decentralization and transport infrastructure of good standards, the figure of more than one car per economically active person is not unrealistic. The ease and convenience of using the right car for the right activity indicates that there is no defined limit to the number of cars per capita.

It has been assumed that as a total the car population ultimately will be about 80% of the number of people living in a country. Sweden in 2040 is still far from this upper limit.

Conclusions

Automobile usage in a future information society will, as illustrated, experience changes. In sum, the new information technologies will not replace travel, but mainly substitute for information carried on paper, but if automotive fuel is rationed in a situation of a supply crisis, some travel will have to be replaced by telecommunications. The synergistic effects of high-quality information will stimulate face-to-face contacts, make new working patterns possible, initiate and stimulate people to new projects, make people interested in visiting new places, and thus stimulate travel and automobile use. A secondary effect of information technology development and improved automobility is that they will make a dispersed living and working pattern possible and desirable. Equally, travel demand and automobile usage will increase as the information society matures. Population growth and longer distances between people in a dispersed society will lead to increasing numbers of passenger kilometres travelled per year. Automobiles will be used predominantly for leisure purposes. Effects of efficient transport pooling with automobiles will thus be marginal.

The number of automobiles will increase and possibly only level off when a car density of one vehicle per economically active person is reached, equalling about 80% of the total population. Utilizing advanced information technologies, future automobiles will be more fuel efficient per passenger kilometre, pollute less, have more automated functions and be more reliable, have systems for better route planning, have much higher safety in operation and have longer lifetimes, move with smoother variations in speed, move with good speed even in dense traffic, be used with higher load factors, be used safely by people without the technical skill required in the 1980s, and be used for a combination of work, business, service, transport, as well as for different leisure activities.

Automobile transport will be available to a larger group of people in the information society. Driving may be permitted from 14 years of age upwards and car pooling can become a social duty. Automobiles will be used more efficiently as a result of better pooling of transport demands. Outside larger cities automobiles with para-transit functions will replace many public transport services. They wil be of medium and larger sizes, be more sophisticated and cost more to buy than automobiles today.

However, life cycle costs and costs per mile driven can be lower. Most will be owned by large companies leasing or renting the cars to the users.

Traffic flow on highways will be safer and smoother; semi-automatic speed control functions can improve traffic flow and decrease travel times, without widening the roads.

The evolving information society is a result of changes in attitudes, values, working patterns, shopping habits and leisure activities, concomitant with advances in technology. Its issues are for the coming generations to decide upon; its new technologies and applications are not yet foreseen, let alone developed. Thus, projections on automobile usage till 2040 are necessarily speculative. However, given the large role that the car will play in future society and the importance of the car industry to the Western economy, the exercise is a necessarily important and interesting one.

Notes

1.   O. Svidén, *Automobile Usage in a Future Information Society*, EKI Transportsystem, Linköping University, 1983.
2.   J. Masuda, *The Information Society as a Post-Industrial Society*, Institute for the Information Society, Tokyo, 1981.
3.   G. Bouladon, "Man, the city and the automobile in the future", *Futures*, February 1974.
4.   N. Calder, "What is futures research?", *New Scientist*, October 1967.
5.   P. Korchelli, "Patterns of urban change", *Options*, IIASA, Winter 1982.

COMPARATIVE ADVANTAGE BY DESIGN

Raphael Kaplinsky

Considering the global manufacturing sector as a whole, rather than specific sectoral components, it is evident that three significant trends have emerged in the post-war distribution of output. First was the rapid rise of Japan as a major industrial power, an advance which saw its share of global manufacturing value-added rise between 1948 and 1980 from 1% to 9.4%; second, an even larger decline in the relative contribution of the USA, whose global share in the same period fell from 56.7% to 21.1%; and finally the rise of the newly-industrializing countries, whose share rose from 3.4% to 7.7%, particularly in South-east Asia. (All figures from UN 1983.)

In the same period, in most of the industrially advanced economies, there has been a significant and protracted trend for the rate of profit to fall and the unemployment rate to grow, as well as for the growth of excess capacity in many manufacturing sectors.[1] The consequence has been a marked sharpening of competitive pressures as individual corporations and national policy formulation strive to restructure production techniques, product lines and organizational frameworks to achieve comparative advantage in world trade.[2] It is in this context (that is, the emergence of "crisis" in the world economy) that we can observe the transition from manufacture to "systemofacture", perhaps as important a phenomenon as the transition over the past few centuries from manufacture to what Marx termed "machinofacture". It is also through this that we can best understand the rapid diffusion of electronics-based automation technologies.

We do not intend to go into a detailed discussion of systemofacture here.[3] Nevertheless it is useful to sketch out its major components, for it is of central importance to our later discussion of the role of design in comparative advantage. Systemofacture involves the evolution of systemic tendencies in contemporary best-practice production environments. Two major types of systemic tendencies are involved. The first is the reorganization of inter-plant relationships, heavily influenced by the organization of production in the Japanese automobile and electronics sectors. This requires the adoption of just-in-time (JIT) manufacturing techniques whereby firms operate on a "zero inventory" basis, a "zero-defect" basis, flexible production schedules and decentralized control over job content. These JIT techiques have key implications for inter-firm relationships such that, optimally, suppliers operate in close proximity to assemblers, and design activities are closely coordinated across plants and firms. The overall result is that the existing pattern of globally widespread, vertically integrated enterprises is likely to be supplanted by geographically proximate and organically linked clusters of production of a systems-like nature.

The second major strand of systemofacture relates to the emergence of intra-firm systems economies.[4] It is in this arena that the new electronics-based automation technologies have particular relevance and where the role of design in comparative advantage is best understood, so it is appropriate that we discuss this in considerably greater detail than the emergence of the inter-firm systemic relationships noted above. This should not be taken to imply, however, that intra-firm systemic tendencies are of greater importance than emerging inter-firm relationships.

The emergence of intra-enterprise systemic automation technologies
The literature in automation has a somewhat muddy history of definitional discussion. At root is the question of whether automation is best understood in cybernetic (i.e. feedback-control) terms, or as a "synonym for advanced mechanization".[5] By differentiating between the emergence at different historical periods of three different elements of automation technology (namely: transfer, transformation and control), Bell[6] makes it clear that the concept is best understood in its wider sense. Yet, useful as Bell's three elements of automation are, his classification is still too narrow to allow us to grasp the full significance of the systemic automation technologies. This is because Bell's analysis is limited to physical manufacture and ignores the development of other, complementary automation technologies such as those in the office and in design. In order to understand the significance of these ancillary developments, it is necessary first to offer a brief description of the evolving pattern of organization of production in the modern enterprise, together with the types of automation technologies used.

In the modern industrial firm, as we can see from Figure 1, there

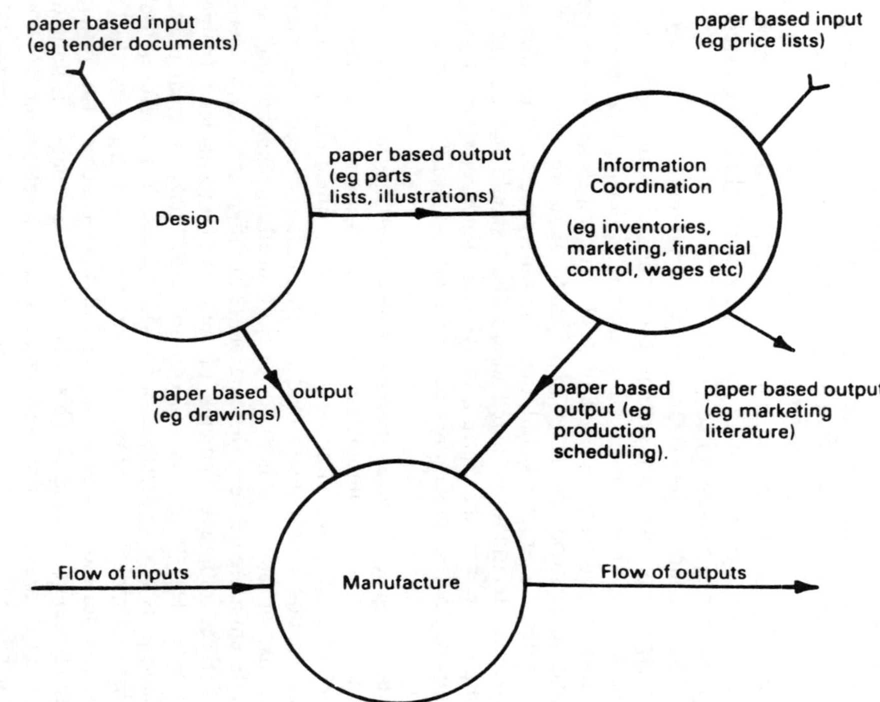

Fig. 1   Pre-electronic organization of factory production

are essentially three spheres of production. The first of these is design, where the nature of the firm's output (e.g. cars, buildings, sweets) is defined and new production processes are explored. The key actors in this sphere are skilled engineers, scientists and technicians, but to work effectively they require the back-up help of a staff of "information processing" assistants such as secretaries and librarians. The actual manufacture of these designs occurs in a second sphere of production in which the raw materials and intermediate inputs are stored, processed into final products and ultimately delivered to the consumer (which is often an affiliate of the same firm). These two spheres of production, which are the kernel of an enterprise's activities, could not operate effectively without some form of coordination, and this comprises the third sphere of production.

Naturally, the extent to which these spheres of production exist in any particular enterprise depends upon the nature of the activity involved. Firms producing simple products with relatively low technology will obviously have a poorly developed design function, whereas small, high-technology electronics firms may have a very well-developed design function which requires little formal coordination. Nevertheless, in almost all modern enterprises, whatever their sector or size, these three spheres of production will tend to be separated into different units (often in different towns, cities and even countries): the R&D block, the factory, and the administration. In some cases, particularly in high-technology sectors, each of these spheres of production may be incorporated in a separate firm. Clearly this separation of function has not always been the predominant form of organization, but has evolved slowly since the onset of the detailed division of labour and the development of manufacture.

Now, within each of these three spheres of production, there is a variety of separate activities. For example, within the design sphere, design itself is usually an activity distinct from drawing, copying and tracing; within the manufacturing sphere there are important differences between handling, forming, assembling, control, storage and distribution; and within the coordination sphere, information has to be gathered, processed, stored, and transmitted. Whilst some activities are common to all enterprises (for example, handling in the manufacturing sphere), there will inevitably be a variation in the number and type of other activities. This variation is particularly marked in the manufacturing sphere, where it will be affected by factors such as the nature of the process (flow or batch) and scale (small or large batch).

Armed with the recognition that there are these three spheres of production, each with its particular sets of activities, it is possible to categorize three different types of automation:

(a) *Intra-activity automation* refers to automation which occurs within a particular activity. Clearly, in line with our earlier definition of automation, this intra-activity automation may take a variety of forms ranging from the simple substitution of machine power for human power (as in the use of computer-aided draughting systems) to the more complex incorporation of machine "intelligence" and control (as in computer-aided design systems). The determining characteristic of this type of automation, however, is that it is limited to a particular activity and that it is consequently isolated from other activities within or beyond the particular sphere of production. In many of the earlier studies of automation, this is referred to as "island" automation.

(b) *Intra-sphere automation* refers to automation technologies which have links with other activities within the same sphere. Indeed, the origins of the term "automation" in the Ford assembly plant of the 1920s illustrate this type of automation well: the new transfer line mechanized the flow of materials between different activities such as lathes, drilling and boring machines. In its more complex form (as in the newly developing flexible manufacturing systems), intra-sphere automation involves the monitoring of the progress of production with an ability to adjust components of individual activities, if this becomes necessary.

(c) *Inter-sphere automation* is the third and most complete form of automation and involves coordination between activities in different spheres of production. In view of the number of activities within each of the different spheres, there is a wide variety of potential inter-sphere combinations. These may be of a relatively limited and simple nature, for example using design parameters to set machine settings automatically; or they may be wide-ranging and complex, such as in the linking of changes in the specification of production to parameters generated in redesign, and thus in continual adjustments made in machine settings.

The essential difference between these three different types of automation is shown in Figure 2. In (a) we illustrate the introduction of automation technologies into individual activities within each of the three spheres. As we can see, there is no link between these individual intra-activity automation technologies and other activties, even within the same sphere. In (b) we illustrate how an automation technology is introduced into a particular sphere

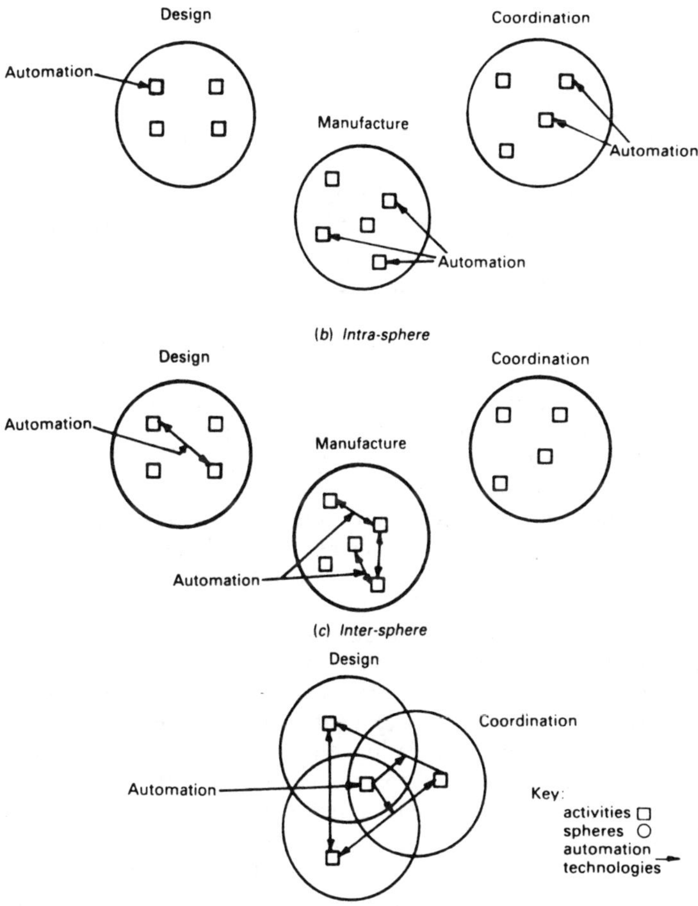
Fig. 2.   The three different types of automation

(a) Intra-activity

Design
Coordination
Automation
Automation
Manufacture
Automation

(b) Intra-sphere
Design
Coordination
Automation
Manufacture
Automation

(c) Inter-sphere
Design
Coordination
Automation
Key:
activities
spheres
automation
technologies
Manufacture

with some form of interlinking (involving feedback in the case of the manufacturing sphere) between different activities. Finally, in (c) we give an example of the merging of the three-sphere industrial enterprise back towards the single-sphere type of organization which, as we shall see, characterized enterprises before the industrial revolution. In this case, automation technology links up different activities between different spheres of production.

By the last quarter of the nineteenth century, the larger enterprises in Western Europe and North America had seen the evolution of the three spheres of production. The sphere of design was increasingly based upon the application of scientific principles, coordination saw the emergence of tiers of management, and manufacture was characterized by the application of ever more complex machinery, involving a steady growth in the division of tasks. The last three-quarters of a century has seen the extension of this differentiated enterprise from the local to national markets, and thereafter from national to international markets.[7] The three spheres of production continued to extend over this period until today we can observe their functioning across the globe. As the extensive literature on the transnational corporation (TNC) shows, design, coordination and manufacture in a single firm often extend over national boundaries: design and senior management in the home country, with manufacture and elements of coordination spread over a great number of countries.

What is at issue now is the transition to the automated enterprise. Whereas the last three centuries have seen the gradual evolution and specialization of the three spheres of production, what we are now beginning to witness is the re-emergence of the unitary, undifferentiated firm. The development of the automated enterprise, embodying the extension of inter-sphere automation throughout the firm, is leading once again to the unity of spheres, as illustrated in Figure 3. This is the significance of drawing out the three types of intra-activity, intra-sphere and inter-sphere automation. Merely focusing on its components (that is, transformation, transfer and control) ignores the central importance of these emerging developments in firm structure and organization.

It is in this transition to the integrated enterprise that the historical significance of microelectronics technology is to be found. There are two reasons for this: the emergence of a pervasive digital (often called "binary") logic, and the dramatic reduction in the cost/capability curve.

(a) Binary systems operate as the basis of either/or logic, in which counting and logical systems can be decomposed into a variety of

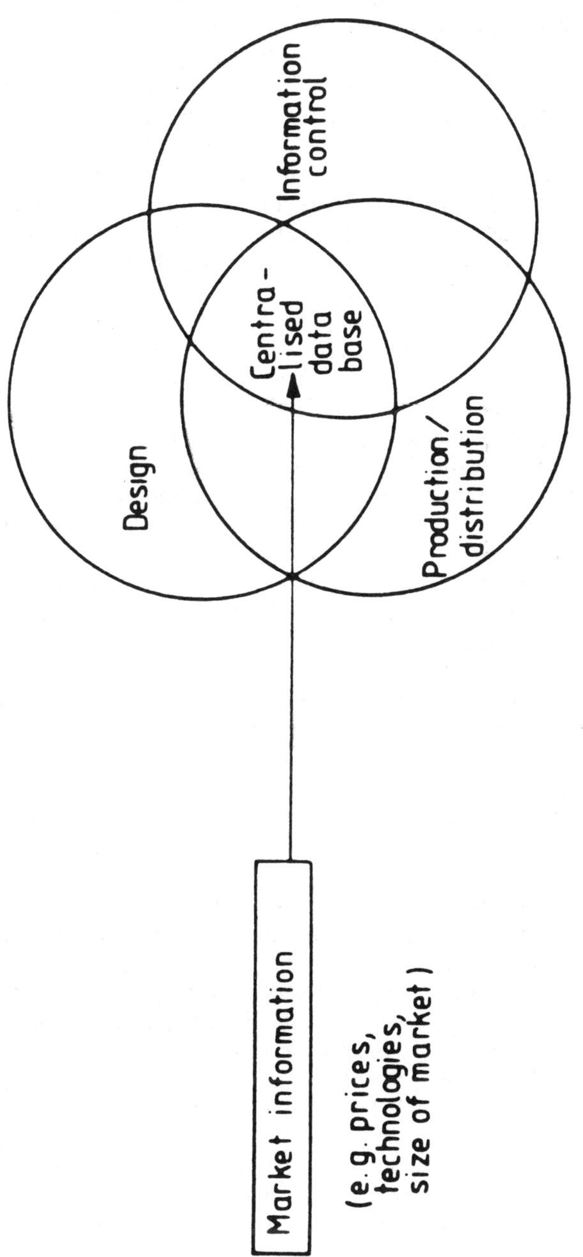

Figure 3   The move to the single system automated factory

stages, each of which can be answered with binary logic. Thus, a common way of *processing* ideas or information can be utilized in a wide variety of activities, across the full range of spheres of production within the enterprise, as well as with external firms and institutions. Since digital logic can be easily transmitted via the interrupted flow of electricity (or light, as is proposed for the future generation of computers), there is a ready interconnection between different digital logic systems. It is this convergence between processing and transmitting information ("informatics") that provides the key facilitating technology for the intra-sphere and inter-sphere automation discussed above.

(b) There is no need to detail the extended decline in the cost/capability curve since the integrated circuit was introduced in 1959. The figures are well known and available in a variety of other sources.[8] However, the significance of this decline is paramount in explaining the pervasiveness as well as the speed with which microelectronics is diffusing in the manufacturing sector.

Hence, in referring back to the three types of automation outlined above, microelectronics in intra-activity automation has tended to be associated with the optimization of control and the storage of information. Indeed, this was the major area of the technology's diffusion in the period between 1960 and the late 1970s. In the manufacturing sphere we saw the maturation of numerical control, beginning with simple machine tools and currently extending to assembly robots; in the design sphere, microelectronics systems began with batch-oriented mainframe design computers and have progressed to interactive computer-aided design (CAD) and computer-aided draughting systems;[9] in the sphere of coordination, applications began with computers being used for stock and wage control, and then extended to word processing and, most recently, to electronic printing.

Then, towards the mid-1970s, the first fledgling attempts were made at intra-sphere automation based upon microelectronics systems,[10] a focus of activity which is currently the major objective of most of the major machinery suppliers providing equipment for each of three spheres of production. In the design sphere, computer-aided design and draughting systems are widely available and have until very recently been the province of new American firms such as Computervision, Calma and Intergraph. In the manufacturing sphere, we are seeing the transition to flexible manufacturing systems, hitherto the speciality of Japanese and Swedish firms with a few exceptions from the USA and Europe. Finally, in the sphere of coordination, a number of firms, predominantly American, are developing integrated multi-function workstations covering the full range of activities.

These efforts at intra-sphere automation, based on digitized electronics technology, represent the cutting edge of technical progress in the mid-1980s. But already the focus of innovative attention in the industrially advanced economies is moving to wider horizons, namely inter-sphere automation. The attraction is to mate together digitized islands of intra-activity automation as well as sets of intra-sphere automation across the three spheres of production. Initially the tasks are narrowly defined, as when CAD systems (in the design sphere) automatically set numerically controlled machine tools in the manufacturing sphere. But the potential is of course far greater, as is beginning to be realized by many major corporations,

engineering firm. GE is restructuring its operations to take advantage of the potential offered by microelectronics to implement inter-sphere automation, both by providing this type of automation technology and by using it. In the former case, GE has in the last four years spent over $700m in acquiring and expanding electronics-based machinery-supplying firms in each of the three spheres of production.[11] With respect to using the new technology, GE has in the same period invested over $2b in re-equipping plants. This includes $316m for a new, inter-sphere automation plant to manufacture locomotives, which offers a good illustration of the potential for systems gains offered by the new electronics-based technology.

Starting at the beginning, the design output of the engineering department will be passed on to the manufacturing engineers in electronic form, rather than as drawings, and will then move through materials control, which will automatically schedule and order materials and keep track of stock and production.

All this information will come together in the factory in the host computer, which will contain in its memory details about how, when and what to produce. This in turn will send instructions to the computer-controlled equipment, such as numerically controlled machines and robots, which will actually do the job. Quality controls, financial data, and customer service records will also be plugged into the same system.[12]

Increasingly, therefore, the direction of technical change in the industrially advanced countries is assuming a *systemic* character, involving the mating together of disparate islands of automation. The organizing thread of this new era of automation is the control of information, the processing and communication of which is vastly facilitated by the utilization of electronically controlled equipment.[13] The significance of the systemic nature of these technological developments should not be underestimated, since it has important consequences for the pattern of innovation. Unlike previous eras of automation where the productivity gains were identifiably related to the introduction of discrete machines, in this new epoch the major productivity gains are being realized when individual sets of equipment in many different parts of the enterprise are linked together. Of necessity this involves two requirements. First is the widespread dispersion of electronics-based equipment throughout the enterprise, and second is the ability to coordinate their workings.

In the former case, there is increasing evidence of widespread diffusion in the industrially advanced economies and in all three spheres of production. Figure 4, for example, charts the rapid spread of electronics-based information-processing automation in the coordination sphere in Japan; Table 1 provides evidence of the spread of numerically controlled lathes in the sphere of manufacture; and Figure 5 charts the very rapid spread of interactive CAD in the design sphere. In each of these cases, innovation was justified in terms of the short-run productivity gains provided in each of separate activities. Yet in historical perspective, the true significance is likely to arise when the individual electronics-based sets of equipment are linked together, as evidenced by the description of the new locomotive plant cited above.

More problematically, the linking together of these items of equipment is constrained not only by technological factors, but also by managerial perspectives, which have historically been based on the decentralization of control and responsibility. There is increasing evidence that electronics-based automated production, built around centralized databases, enables a greater level of control by senior management, but that this requires changes in patterns of managerial organization.[14] Notably, management is required to display a wide enterprise-level of organization, restructuring coordination to take advantage of the potential for systemic productivity gains.

Design in comparative advantage
We begin by summarizing the discussion so far. The post-war period has seen a sharp increase in competitive pressures, associated with changes in the pattern of comparative advantage. In addition to the emergence of Japan as an industrial power, we have witnessed the newly industrializing countries becoming increasingly competitive

Fig. **4**   Percentage of Japanese enterprises using automation technology in the coordination sphere

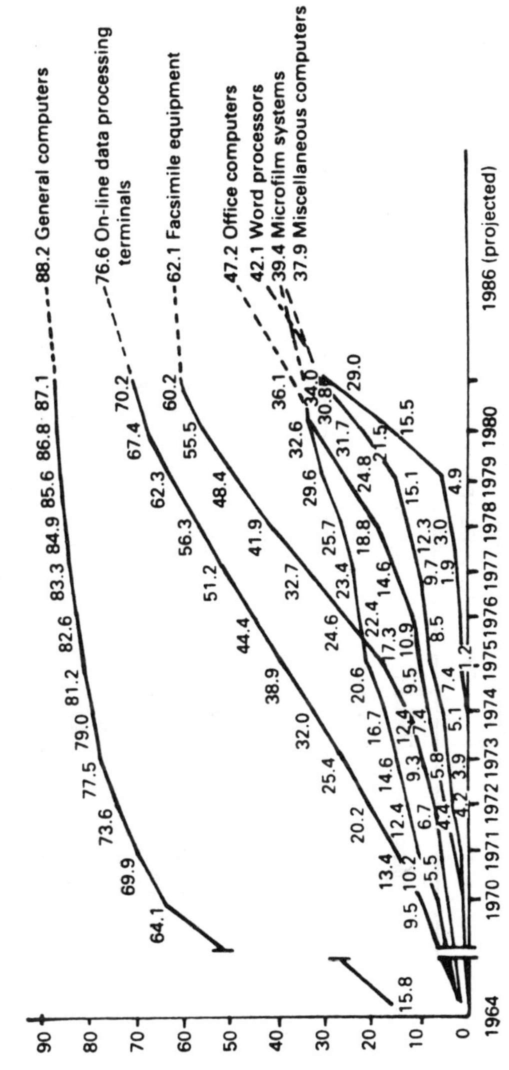

Note: The figures in this chart represent the proportions of firms among 100 enterprises surveyed installing respective types of equipment in the years indicated.

Source: Japan, Foreign Press Center 1982

AUTOMATION IN THE MANUFACTURING SPHERE

Table 1   Investment in NC lathes as percentage of investment in all lathes in major producing countries

Year	United States	Japan	United Kingdom	France	Sweden	Germany, Fed. Rep. of	Italy
1974	n.a.	n.a.	n.a.	n.a.	34.4	n.a.	n.a.
1975	n.a.	23.4	n.a.	n.a.	42.6	16.5	n.a.
1976	n.a.	28.2	18.6	26.4	41.6	n.a.	15.2
1977	n.a.	40.8	21.3	46.7	52.6	n.a.	n.a.
1978	n.a.	40.1	n.a.	n.a.	69.9	n.a.	n.a.
1979	n.a.	50.8	38.4	73.8	69.5	n.a.	n.a.
1980	56.5	n.a.	47.3	n.a.	n.a.	47.1	50.0

Source: UNCTAD 1982c

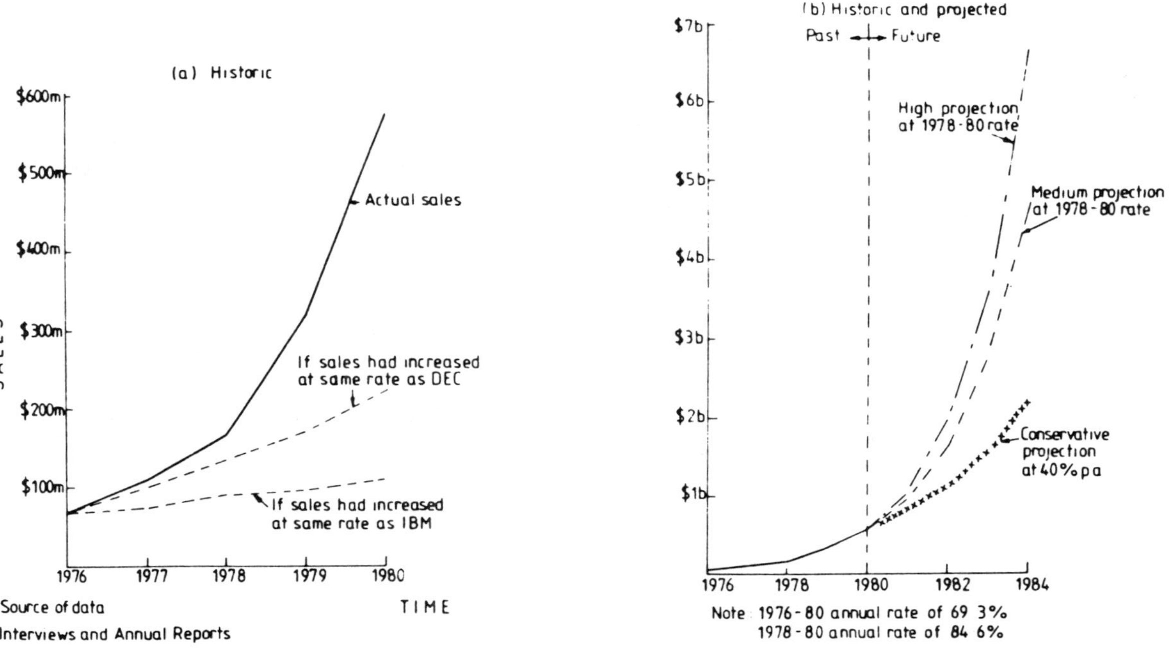

Figure 5   Sales of United States turnkey vendors — past and projected.
(Source: interviews and annual reports)

with the industrially advanced economies. At the same time, and directly related to these developments, we have also seen the emergence of electronics-based automation technologies. The major competitive gains arise when these digital technologies are networked in a systemic way.

Although this discussion has a wide range of implications, the one upon which we will focus our attention concerns the pattern of comparative advantage between the less developed (LDCs) and advanced economies. For some time trade between these two groups of countries has assumed a character which is analogous to that described in conventional economic trade models. In this the LDCs tended to specialize in primary commodities and labour-intensive manufactures; this latter category comprised two bundles of activities, namely labour-intensive goods (e.g. shoes) and labour-intensive operations (e.g. packaging of semi-conductors). The sources of the advanced economies' comparative advantage were a combination of technical knowledge and the skills of a long-established industrial workforce. The particular impact of these skills on comparative advantage varied with the sector; in general they were more pertinent in the discrete-products industries, in the production of capital goods, and when allied to some form of firm-specific comparative advantage.

During the 1960s and 1970s this established pattern of comparative advantage increasingly came to be threatened, and a limited number of newly industrializing countries began to export increasingly technology-intensive manufactures and capital goods, competing directly with their industrially advanced counterparts. This growing competitiveness reflected the depth of their investment in the creation of human resources, the growing skills of their labour forces, and the import of modern technology from the industrially advanced economies. The question now is whether, as projected by the United Nations (which in 1978 set a target for LDCs of 25.7% of global industrial value-added by the year 2000), the comparative advantage of the industrially advanced economies will continue to be eroded in this way.

The diffusion of the new electronics-based automation technologies has an important bearing on this issue, in two divergent ways. If we consider only the diffusion of intra-activity automation technology, then the implication would be a continual enhancement of the LDCs' comparative advantage. This is because the oft observed deskilling tendency of CNC technology[15] is likely to assist LDCs to overcome one of their primary bottlenecks, namely the relative absence of a skilled workforce. For example, to concretize this discussion, we could consider the manufacture of moulds for the plastics industry. In general it takes 10-12 years of experience before a mouldmaker can become proficient; yet with the introduction of CNC machine tools, these skills are becoming redundant; it appears that within five years the skill-profile of the workforce in mould workshops will fall from a high level of intensity to that of mere machine minding. In this case the determinant of comparative advantage becomes the relative wages of the largely unskilled labour force, clearly to the relative advantage of LDCs.

By contrast, if the new technologies are considered in a systemic way, the continued comparative advantage of low-wage LDCs is less evident. For various reasons,[16] we feel that intra-activity automation technologies are more likely to diffuse to LDCs than full inter-sphere automation, thereby allowing the industrially advanced economies to withstand competition from LDCs, and maintain a pattern of comparative advantage which is broadly similar to what currently exists.

Thus, in so far as the industrially advanced economies are concerned, the maintenance of comparative advantage requires the adoption of full inter-sphere automation. And this, as we shall now argue, has particular implications for the role of design in production. To make this point more clearly it is necessary to enter into a brief discussion on the relationship between skills and knowledge, which are related but not identical concepts. Knowledge comprises an understanding of a process of information at an abstract level, such that it can be transmitted to another individual in a similarly abstract manner. As such, knowledge must be explicitly rationalized in abstract terms which can be readily understood — a process which we have come to know as science and technology. On the other hand skill comprises a set of practised experience, which may involve not only the acquisition of knowledge, but also a greater or lesser degree of natural aptitude and implicit rules of operation. Whilst skills are individually acquired and involve a combination of abstract learning, aptitude and experience, the same is not true of knowledge, which is essentially abstract and less individual-specific. Perhaps most significantly, the lesson of the last three hundred years of technical progress has been that knowledge (generally embodied in machinery) has systematically become a substitute for skills. Hence the oft-observed tendency towards the deskilling of work by the substitution of increasingly complex machinery.

Referring back to our earlier discussion of the pattern of comparative advantage in manufacturing in the period since the second world war, in general the industrially advanced countries have had a marked comparative advantage in the skill-intensive and crafts-based industries, especially in the metalworking and capital goods sectors. Whilst this comparative advantage has also arisen from the knowledge inherent in the production processes and product technologies, the long history of artisanal skills has been a major factor in the competitive strength of these industrialized economies. As we have argued, the new, electronics-based automation technologies are having a major impact upon this existing pattern of comparative advantage, since they represent a continuation (albeit at a significantly faster pace) of the trend in which knowledge is being substituted for skill. Indeed the conception of full inter-sphere automation, in which individual machine-settings are all derived from a single centralized database, graphically represents this trend for knowledge to be substituted for skills. But so, too, does the introduction of separate intra-activity components of automation such as CNC machine tools.

However, the introduction of CNC machine tools cannot in itself allow production to take place, for these machines need to be programmed with knowledge. And this knowledge can be created in a variety of different places in the enterprise. The two major loci for creating this knowledge-base are on the shop-floor and in the design office. Which one yields the greatest productivity benefits depends upon whether discrete, intra-activity automation is taking place, or whether the target is wider, intra-sphere or inter-sphere automation.

If individual machine tools are being introduced in a highly selective pattern of automation, then it is probably most appropriate for the machine tool-paths to be calculated at the point of production. In this case, there are no particular implications for the role of design in production. However, if a wider horizon of automation is involved, then logic argues that the design sphere has a key role — if not *the* key role — to play in enabling the firm to innovate productively. This is because, as we have seen, digital logic allows for the productive systemic networking of different electronics-based automation technologies. Since the information base established at the design stage is the one which will subsequently form the core of coordination activities (e.g. parts lists, stock control) and manufacturing activities (notably in machine settings), it is clearly of considerable importance that the initial design be undertaken in a digital electronic format. Moreover, the experience of introducing such design procedures is that the content of the design phase necessarily becomes more science based (that is, knowledge-intensive rather than skill-intensive).

To illustrate this point, let us go back to the example of mould making described above. The firms at the leading edge of technology are not primarily concerned with the introduction of CNC machine tools. Instead they are concentrating upon the digitization of existing designs and the introduction of scientific principles into the design of the moulds (e.g. finite element analysis). Once these tasks have been completed in the design sphere, then tool-paths can be calculated in the CAD system, CNC machines set directly, and inventory controlled automatically. Similar points can be made in other sectors, most notably in the case of group technology and flexible manufacturing systems.

Again, as in the previous discussion, it is possible to respond by questioning whether this represents a new imperative. After all, design has long been recognized as being an important industrial activity. At one level this is a valid comment. Yet to make this observation and at the same time to fail to respond with a significant increase in the effort devoted to design, is to miss the point. The transition to systemofacture involves not just the automation of separate activities within each of the spheres of production. It also involves their interlinking, and without a science-based and systematic policy to design, the unified database[17] which allows for systemic inter-sphere automation will be unobtainable.

Notes

1. R. Kaplinsky, *Automation: The Technology and Society*, Longmans, London, 1984.
2. Trade, as a proportion of output, has shown an inexorable tendency to rise in this same period; see United Nations Conference on Trade and Development (UNCTAD), *Protectionism and Structural Adjustment: An Overview*, TD/B/942, United Nations, Geneva, 1983.
3. R. Kaplinsky, "Electronics-based automation technologies and the onset of systemofacture: some implications for LDC industrialisation", *World Development*, 1984, forthcoming.
4. R. Kaplinsky, *Automation: The Technology and Society*.
5. P. Einzig, *The Economic Consequences of Automation*, Secker and Warburg, London, 1957, p.2.
6. R.M. Bell, *Changing Technology and Manpower Requirements in the Engineering Industry*, Sussex University Press, London, 1972.
7. A.D. Chandler, *The Visible Hand: The Managerial Revolution in American Business*, Harvard University Press, Cambridge, Mass., 1977.
8. See Noyce in T. Forester (ed.), *The Microelectronics Revolution*, Basil Blackwell, Oxford, 1980.
9. R. Kaplinsky, *Computer Aided Design: Electronics, Comparative Advantage and Development*, Frances Pinter, London, 1982.
10. We specifically distinguish these types of intra-sphere automation from the earlier, pre-electronic technologies incorporated in moving production lines.
11. This includes Calma (a leading supplier of CAD equipment), GEISCO (the

world's largest software house), Intersil (manufacturing integrated circuits), SDRA (an engineering services company which assists user firms in introducing automation technologies), and a variety of licences from Japanese and German firms to manufacture industrial robots.

12. R. Lambert, "Geared for a rail revival: General Electric in the US goes for major investment", *Financial Times*, 14 April 1983.

13. One of the more significant consequences of this trend is the emergence of formerly specialized computer hardware firms as comprehensive suppliers of automation technology. IBM, for example, is now the world's largest supplier of CAD equipment and is rapidly moving into the production of industrial robots. Building around its COPICS (communication oriented production information and control system) strategy, they aim to organize production around a unified database.

14. See R. Kaplinsky, *Automation: The Technology and Society*, ch. 7.

15. See: H. Braverman, *Labour and Monopoly Capital: The Degradation of Work in the Twentieth Century*, Monthly Review Press, New York, 1974; D.F. Noble, "Social choice in machine design: the case of automatically controlled machine tools", in A. Zimbalist (ed.), *Case Studies in the Labour Process*, Monthly Review Press, New York, 1979; H. Shaiken, "Computer technology and the relations of power in the workplace", Discussion Paper No. 80-217, International Institute for Comparative Social Research, Berlin, 1980.

16. See R. Kaplinsky, "Electronics-based automation technologies".

17. Which can be handled in either a centralized or a distributed form.

PROSPECTS FOR A NATIONAL INNOVATION POLICY IN THE UNITED STATES

J. David Roessner

Consensus appears to exist on the need for most major industrialized nations to develop and implement some kind of strategically oriented, internally consistent industrial policy. Because these nations recognize the integral role that innovation plays in national economic performance, government industrial policies incorporate, as an integral element, policies intended to influence the pace and direction of technological change in industry. Rothwell and Zegveld (1981) define innovation policy as the fusion of industrial policy and science and technology policy. They believe that it is both necessary and possible for national governments to develop

strategic, long-term innovation policies based on an asessment of current and future technological, economic and social needs and problems, and on an awareness of technological trends and associated commercial opportunities. (p.2)

It is now increasingly recognized that an explicit, national policy for stimulating industrial innovation is an essential and integral part of any government's overall industrial and economic strategy. (p.55)

The United States periodically has initiated policy exercises that might have led to the kind of comprehensive national policy for innovation which Rothwell and Zegveld call for, but these initiatives have come to little. Debate concerning the value and content of *industrial* policies proposed for the USA currently is active and politically visible. What is the likelihood that the USA will develop and implement a comprehensive innovation policy? How does the answer to this question bear on the prospects for a similarly comprehensive industrial policy?

I propose here to examine the histories of US government actions that might loosely be described as efforts to initiate and implement a national innovation policy, to seek explanations for the paths they have taken and for their outcomes, and to infer from the results something about the prospects for a national innovation policy in the United States. My thesis is this: unless a crisis atmosphere (depression, war) prevails for a considerable period (say, as long or longer than a presidental term of four years), the USA will develop and implement neither an innovation policy nor an industrial policy of the comprehensive type described by Rothwell and Zegveld. The reasons for this, I will argue, are deeply rooted in US political values and manifested in its political institutions and policy-making processes. These values have dominated policy making in the USA for more than 200 years; only wartime conditions have significantly altered the way they condition and constrain the policy process.

Although there is an obvious overlap between innovation policy and industrial policy, Merrill's (1984) distinction between the two is useful. Industrial policy, the broader of the two, typically attempts to influence the movement of capital and labour among economic sectors, not only to increase aggregate economic performance but also to mitigate hardships caused by the actions of the market and/or by government. Innovation policy seeks to enhance industrial productivity and international competitiveness by encouraging technological change; it usually omits concern for equity. Innovation policies typically target capital formation, R&D investment, technical manpower training, promotion of technology diffusion, and protection of intellectual property; industrial policies may include all these plus capital allocation, trade, and labour policies. Both types of policies may employ procurement, regulation, and taxation. The greater breadth of industrial policy is one reason its history in the USA reads so differently from that of innovation policy.

History of industrial policy and innovation policy in the USA
The United States has never had an industrial policy or an innovation policy in the strategic, comprehensive sense. US industrial "policy" has been a series of government subsidies, tax benefits, tariffs, and economic regulations, each enacted in response to a clientele or interest group (small businesses), industry (civil aviation), or short-term situation (inflation), with little or no consideration of their impact on, or interaction with, other policies affecting industry. Recently economic conditions in the USA have generated much comment on this situation. Harlan Cleveland (1982) regards this historical pattern as pernicious and in need of change:

Ad hoc tactics by the federal government — rescuing Chrysler, suing IBM and AT&T, embargoing selective foods and feeds, and jawboning the Japanese — serve both as evidence that we don't have, and that we badly need, a "strategic" sense as a nation of where we are trying to go, what we are trying to do. (p.vii)

In another observer's view, the legislative, judicial, regulatory and executive apparatuses of government in the United States have established policies more in conflict than in harmony. "Sometimes it would appear as though they were giant wrecking machines, flailing at one another across the Mall" (de Simone, 1982, p.119). Mansfield can offer no good economic reason why the US government spent so much money on civilian aviation technology and so little on railroad technology, or why so much on agriculture and so little on building construction (Mansfield, 1976, cited in Wescott, 1983). Wescott comments on the three major "bailouts" of industrial firms that took place during the 1970s:

Because of the closeness of the congressional votes it seems appropriate to conclude that the word *response* is probably more accurate than the word *policy* as far as US bailouts are concerned. (p.104)

In sum, US industrial policy results largely from an overall commitment to the efficiency of the marketplace and to *laissez faire* ideology, amended by periodic, uncoordinated, and narrowly conceived interventions responsive to short-term political exigencies.

The history of innovation "policy" in the United States reads much the same. The US government has always had innovation-related policies or programmes, despite the fact that explicit attention to innovation as one key to economic performance is a recent phenomenon. Rooted in the Constitution itself, the Patent Act of 1790 acknowledged the value to the nation of stimulating the rate of invention by granting inventors temporary monopoly control over the sale of the physical embodiment of their ideas. Beginning with the Hatch Act of 1887, the federal government and the states created a vast and highly effective cooperative agriculture research system, extending from the Department of Agriculture through Land Grant Colleges to County Agents. The primary purpose of the system was to stimulate agricultural productivity through the development of new seeds, equipment, fertilizers, and farm practices. During the first half of the twentieth century, government intervened in the private economy in myriad ways to promote or shape innovation. For example, the National Advisory Committee on Aeronautics (founded in 1915) conducted engineering tests that speeded the pace of innovation in commercial aviation; what was to become a massive federal investment in biomedical research began in the 1930s with the National Cancer Institute; in 1954 the Internal Revenue Service implemented a rule allowing businesses to treat research and development expenditures as current busines expenses for tax purposes. These policy measures were piecemeal and ad hoc, usually focused on a single industry or scientific field and rarely integrated with broader economic policy. Policies intended to influence the pace and direction of technological innovation were, and still are, individualized responses to groups, industries, and economic sectors.

If one thinks of the policy-making process as a series of stages, e.g. issue aggregation, issue formulation, agenda-setting, policy/ program design, legitimation, and implementation, it is clear that *individual* actions intended to affect industrial innovation passed successfully through each stage of the process. But beginning in the 1960s, coincident with the emerging consensus among economists that technological change in industry was closely linked to economic performance, the United States launched a number of programmes and policy exercises whose original scope and objectives more closely resemble the current definition of innovation policy. In most cases, however, these initiatives (as originally conceived) never reached policy-makers' agendas.

J. Herbert Hollomon, Assistant Secretary of Commerce for Science and Technology during the Kennedy administration, proposed a Civilian Industrial Technology Program in 1963, intended to promote innovation in what were regarded as technologically backward industries such as building construction and textiles. In 1965 he implmented the State Technical Services Program, modelled after agriculture's Cooperative Extension Service, which would provide university-based R&D and technical assistance to businesses. Congress approved only the State Technical Services Program, but eliminated funding for it in 1969.

In 1971, under the New Technological Opportunities Program, President Nixon attempted to identify technology-based initiatives which government could undertake to improve the nation's deteriorating economic situation. Agencies submitted some $11 billion in project ideas to the White House, but all that emerged were $40 million in proposed programmes intended to increase understanding of how government actions affected industrial R&D. OMB impounded most of these funds almost immediately.

President Carter's Domestic Policy Review (DPR) on Industrial Innovation, begun in 1978, involved an elaborate network of agency task forces, research contractors, and advisory committees with business, academic, labour, and public interest representatives.

Dozens of recommendations were submitted, but internal reviews screened out many of the advisory committee's highest priority items such as tax incentives for capital formation and for investment in R&D and small, high technology companies, major regulatory reforms, and changes in trade and export policies. White House recommendations were modest, amounting to $55 million in proposed programme expenditures. One of the centrepieces, the $5.2 million Cooperative Generic Technology Centers Program, was deleted from the budget by the Reagan administration just over a year later.

Innovation policy in the strategic sense has never reached the political agenda. Literally hundreds of recommendations for government action to promote innovation have been promulgated, most of them developed within the context of some major policy exercise such as the DPR (Mogee and Kremer, 1980). What emerged from the policy process were mere shadows of what entered. Proposals were elminated, weakened, or compromised so that the results were narrowly conceived and poorly integrated. In the case of the innovation policy initiatives just described, none of the proposals enacted lasted beyond the initiating administration's term of office. Several features characterized the political process which surrounded these initiatives: industry representatives did not instigate them and in some cases were antagonistic to them; proposals for action originated with the executive branch, with Congress playing a negligible role; only narrowly based interest groups were involved (academics, the industrial research community, OMB, and the Treasury Department) (Merrill, 1984).

The recent flurry of action surrounding industrial policy stands in contrast to this experience in several respects. Industrial policy has gained a prominent place on the political agenda. Numerous hearings were held and more than 30 bills were introduced in the 98th Congress. It is a major topic in economic policy debates which involve key constituencies in American society. Rather than bipartisan and academic in flavour, as was the case with innovation policy initiatives, industrial policy has been partisan and politican from the outset (Merrill, 1984). What explains the abortive nature of US innovation policy initiatives?

Some possible explanations

When stated in its most general form, innovation policy appears to benefit everyone. Who could oppose improved performance of the US economy resulting from increased innovation? But with specific innovation policy proposals come groups who recognize whether they will win or lose if the proposals are implemented. The Civilian Industrial Technology Program was intended to stimulate R&D in building construction. A cement industry representative recognized that "if we sponsor research for basementless slabs, we don't get the concrete that goes into the foundation" (Nelkin, 1971, cited in Mogee, 1984). Industrial R&D executives and managers compete for resources with directors of other corporate functions. The responses of various interest groups to the Carter DPR proposals illustrated the absence of consensus. As Mogee (1984) notes: "Policy changes recommended by industry were characterized as 'regulatory rollbacks,' 'tax breaks,' and 'weakened antitrust protections' by the other participants." Bean and Baker's (1984) tabulation of public interest group reactions to the nine (relatively innocuous) proposals that emerged from the DPR exercise and to provisions of Reagan's 1981 Economic Recovery Tax Act shows a mixed reaction, at best (Table 1). Mogee (1984) has examined the reasons a recent White House initated study failed to produce detailed, specific recommendations as the Office of Science and Technology Policy had requested. *America's Competitive Challenge: The Need for a National Response*, provided good public relations and legitimated administration proposals, but could not develop concrete recommendations that could be translated easily into government action. The members of the study's advisory committee, chief executive officers of business firms and universities, were unable to agree on specific proposals. According to Mogee, the explanations for this were two-fold: the fundamental differences in values and interests between universities and business, and the failure of participants to devote sufficient time and resources to generate informed debate. This illustrates the conflict that inevitably arises when interest groups participate in innovation policy formulation in the United States, and the relatively low priority that universities and businesses place on innovation policy.

Merrill (1984) has analysed the processes by which US innovation policy initiatives were developed. He concludes that

Dispersed executive branch and congressional jurisdiction over tax, regulatory, R&D, and other policies of concern to particular constituencies ... handicaps creation of a broad, stable coalition, even with presidential leadership.

According to Merrill, innovation policy (in contrast to industrial policy) has failed to command a place on the political agenda because it has failed to engage politically active and powerful

Table 1

R&D Activities Potentially Impacted by
Industrial Innovation Policy Initiatives

<u>Type of R&D Activity</u>	<u>Public Interest Position</u>
Relations outside the firm	
a) R&D/technical cooperation	
1. Firm/firm	Opposes relaxation of antitrust; supports clarification of practice
2. Firm/university	Supports some elements, neutral on others
b) Technical information and property rights	
1. Government acquisition of information of foreign technologies	Opposes
2. Uniform government patent policies	Supports
3. Transfer of government-owned patents to industry	Opposes
Decisions and activities inside the firm	
a) R&D budgets and strategies	
1. Overall R&D budget increases	Mildly supports
2. Develop internally/acquire outside	Pro small business
3. Accelerated cost recovery of new R&D plant and equipment	Mildly opposes
b) R&D project selection and portfolio structure	
1. Decrease regulatory compliance/ defensive research	Strongly opposes
2. Projects targeted at government markets	Supports
3. Accelerated capital cost recovery for investment in new R&D plant and equipment	Mildly opposes
c) R&D personnel/human resources	
1. Compensation/awards for employed investors	Supports increases
2. Revision of tax reg. I.861-8	Supports
3. Facilitation of labour adjustments to technological change	Supports

(Source: Bean and Baker, 1984)

constituencies and because of the corresponding roles of Congress and the bureaucracy. The decentralized structure of Congress meant that support was forthcoming only for those programmes that fell within a narrow committee jurisdiction, e.g. COGENT and technology transfer proposals. Executive agencies involved in developing innovation policies were politically weak (commerce) and similarly dispersed. In contrast, industrial policy was seen as inseparable from the fates of major US industries, thus commanding the attention of groups representing large employers, their associated unions, and the nation's security and strength. Merrill neatly summarizes the irony:

characteristics associated with the political failure of national innovation policy in the United States are widely regarded as conditions for its effectiveness — lack of partisanship, ideological cleavage, and high visibility that are likely to undermine policy stability.

Political stability matters only after an issue gets on the political agenda and interest group concern is translated into authoritative government action.

Conclusions

In my view, the reason innovation policy has failed to gain recognition among US policymakers is the same as the reason that industrial policy proposals will fail following the current flurry of attention they are receiving. It has to do with the fundamental underpinnings of the US political system. In response to a deep-seated fear of tyranny by either a majority or by minority factions, the framers of the US Constitution fashioned a political structure that features the most powerful popularly elected legislature in the world, political authority shared among three branches of national government, and a federal system characterized by multiple centres of political authority independent of the national government. Among modern industrial nations, US mechanisms for achieving coordination in policy making represent an extreme on a continuum from central supervision to negotiation, bargaining, and other forms of mutual adjustment (Lindblom, 1968, p.82). Lindblom is an apologist for the "inefficiency" of pluralist policymaking:

Because in a pluralist, complex society, social goals are not a tightly knit harmonious structure, because we value openness in goal structure, and because we keep the social peace by permitting conflicting groups to pursue conflicting goals, we do not even want to reject such apparent inconsistencies as — to take a common example — subsidizing some farmers to restrict output, and others to expand it. Depending on who the farmers are, what their crops are, where their land is, and how subsidies affect them, both subsidies may make sense. (p.109)

Political scientist Robert Dahl (1956) identifies the highly participative, fragmented nature of the American political system as perhaps its most distinctive feature.

A central guiding thread of American constitutional development has been the evolution of a political system in which all the active and legitimate groups in the population can make themselves heard at some crucial stage in the process of decision.

It may be that in the case of innovation policy the United States cannot "afford" the luxuries of widespread public participation, decentralized decision making, and extensive bargaining and compromise that must occur before government acts. Affordable or not, these values are too deeply embedded in the American social fabric and its political institutions to change. In the USA, neither innovation policy nor industrial policy can be formulated in "small circles of officials and experts", as apparently is the case in Europe (Rothwell and Zegveld, 1981, p.236). Short of a crisis, there will be no comprehensive innovation policy or industrial policy in the United States in the foreseeable future.

References

Bean, A.S. and Baker, N.R. (1984)
"Implementing national innovation policies through private decisionmaking", in J.D. Roessner (ed.), *Government Innovation Policy: Design, Implementation, Evaluation*, Port Washington, NY, Associated Faculty Press.

Cleveland, H. (1982)
"Vitalization without the 're'", in M. Dewar (ed.), *Industry Vitalization: Toward a National Industrial Policy*, New York, Pergamon.

Dahl, R. (1956)
A Preface to Democratic Theory, University of Chicago Press.

deSimone, D. (1982)
"Government and innovation", in S. Lundsted and W. Colglazier (eds), *Managing Innovation: The Social Dimension of Creativity*, New York, Pergamon.

Lindblom, C.E. (1968)
The Policy Making Process, Englewood Cliffs, NJ, Prentice-Hall.

Mansfield, E. (1976)
"Federal support of R&D activities in the private sector", in *Priorities and Efficiency in Federal Research and Development*, papers submitted to the Joint Economic Committee, Washington DC, US Govt Printing Office.

Merrill, S. (1984)
"The politics of micropolicy: innovation and industrial policy in the United States", *Policy Studies Review*, May.

Mogee, M.E. (1984)
  "Knowledge and politics in innovation policy design", *Policy Studies Review*, May.
Mogee, M.E. and Kremer, R. (1980)
  "Two decades of research on innovation: selected studies of current relevance", in US Congress, Joint Economic Committee, Special Study of Economic Change, vol.3, *Research and Innovation: Developing a Dynamic Nation*, Studies, 96th Cong. 2nd Sess., Washington DC, US Govt Printing Office.
Nelkin, D. (1971)
  The Politics of Housing Innovation, Ithaca, NY, Cornell University Press.
Rothwell, R. and Zegveld, W. (1981)
  Industrial Innovation and Public Policy, Westport, CT, Greenwood Press.
Wescott, R.F. (1983)
  "US approaches to industrial policy" in F.G. Adams and L.R. Klein (eds), *Industrial Policies for Growth and Competitiveness*, Lexington, MA, Heath.

GOVERNMENT POLICY, TECHNICAL CHANGE, AND INDUSTRIAL STRUCTURE: THE US AND JAPANESE COMMERCIAL AIRCRAFT INDUSTRIES, 1945-1983

David Mowery and Nathan Rosenberg

The Japanese aircraft industry has attracted considerable comment in articles and public documents published recently in the United States. In many cases the Japanese industry has been cited as a threat to the future dominance of world markets by the US commercial aircraft industry. The following statement by the General Accounting Office is representative:[1]

> Many U.S. Government and industry representatives believe that Japan can and eventually will become a serious competitor in the world's civil aircraft market; the remaining questions are when, and how much Japan's share will be ...
>
> While Japan's aircraft industry is small by comparison with its U.S. and European counterparts, it is expanding and being encouraged with government support.

This paper examines the historical and prospective development of the Japanese commercial aircraft industry. We are particularly interested in comparing the patterns of government policy and industrial development that have characterized commercial aircraft in the United States and Japan during the post-war period. The comparison of commercial aircraft with previous Japanese industrial export successes is another topic of interest. In both areas of comparison (i.e. the USA and Japan, and commercial aircraft and other industries within Japan), some novel conclusions emerge. The US commercial aircraft industry has been a major beneficiary of government policies which (prior to 1978) simultaneously affected the demand for and the supply of technological innovations. In many ways the pre-1978 policies towards the US aircraft industry resemble those observed in such Japanese industries as computers or semiconductors. This unique position of this industry reflected the close technological and financial links between the military and commercial aircraft industries.

The comparison of commercial aircraft with other Japanese export industries suggests that fears of the imminent Japanese competitive threat in aircraft are somewhat exaggerated. The character of the global and Japanese domestic markets for commercial aircraft are such that a capacity within Japan for the independent design, assembly and sale of large commercial transports is not likely to be achieved prior to the 21st century. This forecast is based on several factors. The nature of the domestic market for commercial aircraft is such that a "catch-up" policy (i.e. one emphasizing the importation of foreign technology for initial application in a more or less protected home market) is not feasible for Japan. Moreover, the research and engineering infrastructure necessary to support an independent technical capability in this industry is also lacking within Japan at present.

The analysis of the Japanese and US aircraft industries raises a number of more general issues of interest, including the operation and organization of multinational joint ventures. It is far from obvious, for example, that such joint ventures function as channels for the effective transfer of key technologies. The rapid growth of multinational joint ventures in product development in this and other industries (e.g. the five-nation V2500 jet engine project) also points to the need to reformulate the conceptual framework for the analysis of multinational firms and investment activity.

We begin below by briefly surveying the economic and technological significance of the US commercial aircraft industry. This section is followed by a description of the historical development of the Japanese aircraft industry, its current structure, and key features of government policy towards the industry. Government policy and industrial development in the US commercial aircraft are discussed in the next section, to provide a basis for comparison with Japan. The prospects for the Japanese industry, and associated issues for consideration by Japanese and US policy-makers, are considered in the conclusion.

The economic and technological significance of the US commercial aircraft industry
As a high technology success story within the US economy, the commercial aircraft industry has few peers. Commercial aircraft have experienced rapid rates of technological change. The rate of productivity growth in US commercial air transportation, the primary industrial beneficiary of technological improvement in commercial aircraft, has been equalled or surpassed only by that of the telecommunication services industry.[2]

Within the US economy, the commercial aircraft and aerospace industries are of considerable significance. In 1982, total sales of the aerospace industry (including missiles and spacecraft, in addition to aircraft) amounted to nearly $74 billion, more than 2% of total Gross National Product (GNP). Within this total, sales of military and civilian aircraft, engines and parts were valued at $32.6 billion.[3] Military sales accounted for slightly more than 50% of this total in 1982. The contribution of aircraft to US foreign trade in 1982 was also of considerable importance; aircraft exports equalled $15.6 billion, the largest single category of manufactured exports.

The US aircraft industry also plays a major role in extending and influencing the technological underpinnings of the economy. The aircraft industry is a major investor in research and development. R&D expenditures (74% of which were financed by federal funds in 1978) amounted to 12.4% of the value of 1977 shipments, a level exceeded only by the electronics industry. In addition, the aircraft industry has important linkages (i.e. through its demand for components and parts) with other high technology sectors of the US economy.[4] The linkage between aircraft and electronics is the most obvious (e.g. avionics and computer-aided design and manufacturing), but the aircraft industry also contributes to the support of sophisticated materials development and fabrication activities. Indeed, a key factor in the rapid technological progress that has characterized aircraft is precisely the ability of the indsutry to darw on and benefit from technological developments in a wide range of other industries.

The technological and other links between the aircraft industry and a wide range of seemingly unrelated industries reflects an important aspect of the industry, namely, the high degree of systemic complexity embodied in its products.[5] An individual aircraft is composed of numerous components and subsystems responsible for propulsion, navigation, etc. which are each very complex. The interaction of these individually complex systems is crucial to the performance of an aircraft, yet is often extremely difficult to predict from design and engineering data. Uncertainty about aircraft performance is also exacerbated by the still modest state of scientific theory concerning the behaviour of such critical components and materials. A substantial element of technological uncertainty thus exists in the design and production of new aircraft.

The dependence of commercial aircraft on technologies and components from other industries is reflected in the importance within this industry of a large population of small vendors and subcontractor firms, engaged in production and components supply for the much smaller group of major contractors (the growing importance of subcontracting and the reasons for this growth are discussed below). Estimates of the size of this population of supplier firms within the United States range up to 15,000. The substantial specialized supplier network is an important factor in the historical self-sufficiency of the US commercial aircraft industry. As is discussed below, it is this vendor network, rather than the major commercial aircraft producers, that is likely to bear the brunt of Japanese competition in aircraft.

Key aspects of the current technological and market environment
In order to understand the current prospects for the US and Japanese commercial aircraft industries, it is necessary to examine several aspects of the technology and market prospects for large commercial aircraft. Among the most important features of the economic environment of this industry is the dramatic rise in the costs of new product development for both engines and airframes. Figure 1 depicts the growth in development costs (adjusted for aircraft size and excluding the engine) from 1930 to 1970, a period that began with the introduction of the Douglas DC-1 and ended with the McDonnell Douglas DC-10 and the Boeing 747. In constant dollar terms, size-adjusted development costs have risen at an average annual rate of nearly 20%, far outstripping the growth in aircraft weight (an average annual rate of 8.5%). Miller and Sawers[6] estimate the development costs of the Douglas DC-2 at roughly $150,000, and slightly more than $3 million for the DC-3. The arrival of the jet transport dramatically increased these costs; the DC-8 required $112 million, while the Boeing 747 ran to nearly $1 billion. More recently, the development of the Boeing 767 has been estimated to have cost $1.5 billion, while the proposed Boeing 150-seat jet aircraft is expected to require at least $2 billion for development. Similarly spectacular figures are cited for the costs of development of new engines. The recently announced V2500, a high bypass engine designed to produce 23-25,000 pounds of thrust, is widely expected to cost $1.5 billion for development.[7]

One result of such sharp increases in the fixed costs of product innovation in this industry has been an escalation in the financial risks of innovation. This increasingly risky environment has contributed to exit from the US commercial transport industry (Lockheed and, arguably, McDonnell Douglas having withdrawn from significant new product development and manufacturing activities in the past three years). Producers have also attempted to spread such risks, through subcontracting and (more recently) joint

TRANSPORT AIRCRAFT DEVELOPMENT COST TRENDS

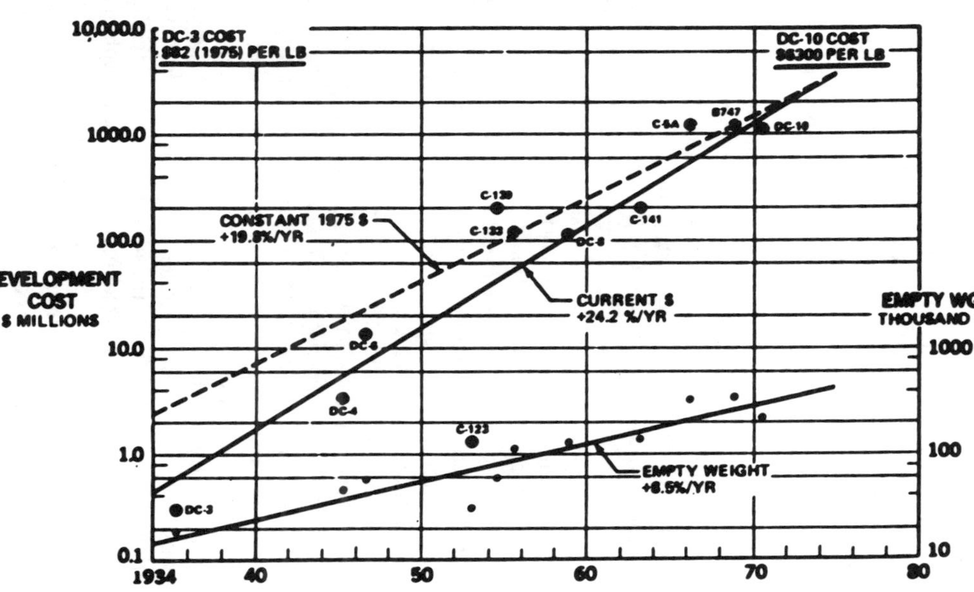

SOURCE: DOD-NASA-DOT STUDY "RADCAP" AUGUST 1972

Source: McDonnell Douglas Corporation, Douglas Aircraft Company, Office of Planning.

ventures in product development.[8]

Another development in the market for commercial aircraft that is of particular importance for American producers concerns the declining share of world aircraft demand accounted for by the United States. In 1971, the United States accounted for roughly 57% of the total air miles flown in the non-communist world, while US airlines purchased 67% of the aircraft produced by US firms during the period 1950-1970. However, after 1970, the rate of growth in the US air transport market slowed and other regions now account for a significantly larger share (notably the industrializing nations of East and Southeast Asia) than previously. From 1977 to 1982, only 40% of all orders for new commercial aircraft were placed by US airlines. While the US still represents the largest single market, it no longer constitutes an absolute majority of world demand. Indeed, in the face of mushrooming development costs, it is unlikely that a new commercial aircraft can achieve financial success without substantial foreign sales. Penetration of foreign markets, while always important to US firms, now is essential. Foreign marketing prospects in many cases are enhanced greatly by allowing local manufacture of some portion of a new aircraft or its engines. The increasingly international character of the aircraft demand, by creating an incentive for producers to offer local "offsets" in the manufacture of aircraft or components, thus works to increase international cooperation in commercial aircraft development, production and sales.

The outlook for the world aircraft industry is also affected by technological developments. In the near future (i.e. prior to the 21st century) technological developments are likely to have a great impact on design and production processes for all aircraft components: airframes, engines, avionics, and materials. The development of computer-aided design and manufacturing technologies (CAD/CAM) has significantly improved the ability of major firms to subcontract or otherwise decentralize design and construction activities. In effect, the ability to transfer technologies in intermediate stages of development is greatly enhanced by CAD/CAM. Manufacturing processes, as well as aircraft design and performance, will be significantly altered by increasing employment of non-metallic composite materials in primary structures (e.g. wings or fuselage) of aircraft. Composites, which utilize such materials as boron or carbon fibre, have the potential for significant reductions in aircraft weight. Currently, secondary structures (such as 767 rudders) are fabricated from composite materials. Employment of composites in major primary structures, however, will require novel manufacturing processes, centred around the cutting, layering and gluing of clothlike materials, in sharp contrast to current techniques which emphasize the bending, joining and grinding of metallic materials. The technical details of aircraft design are also likely to require major revision. The extensive use of composite materials in large commercial transports thus appears unlikely prior to the year 2000.

Through much of the post-war period, the development of new transport aircraft has been dependent on the introduction of new engines. The large high-bypass engine technology, developed for the US C-5A military transport, led to the 747. Subsequent new aircraft have exploited the development of progressively smaller high-bypass engines, conferring the economies and other performance advantages of this engine technology upon smaller aircraft. The latest version of the Boeing 737 employs newly developed high-bypass engines generating roughly 20,000 pounds of thrust. The prospective 150-seat transport is also closely linked to the development of an appropriate high-bypass engine (e.g. V2500). However it appears that the prospects for significant product innovation in airframes (including control and aerodynamics), as opposed to engines, will provide the most significant improvements in performance as well as operating costs for commercial aircraft.

Another area of significant technical advance is that of avionics, including the exploitation of fibre optics, digital communications, and the vast increases in computing power and miniaturization made possible by microprocessors. However, avionics is an area of aircraft technology that lends itself much more to retrofitting of existing aircraft (as, to a lesser extent, do engines). While new commercial aircraft are able to exploit the state of the art in avionics, technical progress in this area is not likely to influence the pace and form of product innovation to the same extent as materials or engines.

The development of the Japanese aircraft industry 1945-75
Having briefly discussed the current outlook for the market and technologies associated with large commercial transports, we now turn to a brief historical overview of the development of the Japanese aircraft industry since 1945. Perhaps the most critical influence on this evolution is the complex relationship between the military and civilian segments of the industry, which is typical of the aircraft industry generally. What makes the Japanese case unique is the fact that this relationship has involved both the American and the Japanese military. This unusual status of the

Japanese aircraft industry has also meant that the Ministry of International Trade and Industry (MITI) historically has not been the sole or most powerful governmental actor. Some observers have also argued that the close ties between the military and civil segments of the industry have reduced somewhat the ferocity of competition among Japanese aircraft firms.

Prior to and throughout World War II, the Japanese aircraft industry was both large and technically sophisticated. During the peak period of wartime production, the Japanese industry was producing 25,000 airframes and 40,000 engines per year. In the aftermath of the war, the aircraft industry was completely dismantled by the occupation forces, and the production of aircraft was prohibited.

The outbreak of the Korean War, followed shortly by the end of military occupation, led to the revival of the Japanese aircraft industry. The Korean War resulted in a vast expansion in military aircraft repair and service activities, areas in which Japanese firms became active with the encouragement of US military authorities. Following this early stage of reconstruction, the end of military occupation in 1952 effectively removed the prohibition on aircraft manufacture in Japan. In 1954, the first Aircraft Promotion Law, allowing for collaboration among firms on aircraft development and providing liberalized investment and depreciation credits, was passed. This was also the year in which the Mutual Defense Assistance Agreement between the United States and Japan was implemented.

An important component of this treaty was the provision for Japanese production of US military aircraft for use by the Japanese self-defence forces. Beginning in the mid-1950s, when Mitsubishi Heavy Industries began production of the F-86 fighter and the T-33, Japanese production of US aircraft has continued to the present, with Mitsubishi currently involved in the production of the F-15. Kawasaki Heavy Industries currently manufactures the P-3C patrol aircraft and the Chinook helicopter under co-production agreements, while Ishikawajima-Harima Heavy Industries is manufacturing the Pratt and Whitney jet engines for the F-15. Other military aircraft manufactured on a co-production basis in Japan include the F-104 and F-4 fighter aircraft.

Typically, Japanese involvement in co-production began as the assembly of kits of US-produced components and subassemblies. Over the life of each co-production agreement, however, the Japanese content of the aircraft has risen (the F-15 is an example of this). Clearly, the US-Japanese co-production agreements served to transfer considerable production and technical knowledge to the major Japanese aircraft firms. However, it is not clear that the *design* capabilities of these firms have benefitted appreciably, nor is it the case that much of the technology employed in fighter aircraft is readily transferable to applications in commercial aircraft. Nonetheless, the fact that aircraft produced under licence in Japan may frequently cost as much as 100% more than identical aircraft produced in the USA (as in the case of the F-15, a difference that is due to the small size of Japanese production runs) suggests the possibility that one motive for Japanese participation in these co-production agreements is precisely the development of a stronger Japanese aircraft industry.

The second Aircraft Promotion Law, passed in 1958, established the Nippon Aircraft Manufacturing Company and laid the foundations for the development of the YS-11, the first Japanese commercial transport of the post-war era. The law and the YS-11 were advocated by MITI, and 54% of the funds for the design and development of the YS-11 were provided by the government. The NAMC was a consortium, including Mitsubishi Heavy Industries, Kawasaki Heavy Industries, Fuji Heavy Industires, Showa Aircraft, Japan Aircraft, and Shin Meiwa Industries. The aircraft, a 64-passenger, two-engine turboprop, was designed for relatively short slights and short takeoffs and landings. As such, the aircraft was well suited to the needs of the domestic Japanese market, and roughly 120 were sold to Japanese airlines.

However, foreign sales of the YS-11 failed to meet expectations, due to a number of factors. The sales and marketing network of the NAMC consortium was modest, and product support activities (maintenance, parts and repair services) were weak. In addition, the commuter market in the United States in particular was relatively small, due in large part to the regulatory efforts of the Civil Aeronautics Board. The YS-11 also faced strong competition in the commuter airline market in the USA and Europe from the Fokker F-27. Despite exports of roughly 60 aircraft out of a total porduction history of 182, the YS-11 was a financial failure. The Japanese home market was too small to support the profitable introduction and sales of an aircraft designed primarily for that region. Without significant exports, Japanese aircraft could not achieve commercial success. The NAMC was dissolved in 1982.

In the late 1960s and early 1970s, the major Japanese aircraft firms (primarily Mitsubishi and Kawasaki Heavy Industries) began to expand their activities as subcontractors for major US commercial aircraft producers, producing small assemblies and components for the Boeing 747 and the McDonnell Douglas DC-9 and DC-10. In an effort to encourage a more extensive and sophisticated international cooperation between Japanese and foreign aircraft manufacturers, the Japan Commercial Transport Development Corporation was formed under MITI's sponsorship (participant firms were Mitsubishi, Kawasaki, and Fuji Heavy Industries) in 1973. Preliminary design and development work began on what was known as the YX; negotiations with the Boeing Company culminated in a Memorandum of Understanding in 1978, in which the JCTDC was committed to produce 15% of the fuselage of the Boeing 767.

Public support of the JCTDC primarily took the form of *hojokin*, low-interest loans that are repayable upon the realization of a profit. Such financial assistance covered 50% of the early design and development costs of the YX programme. Eckelmann and Davis[9] estimate that these loans amounted to $72.6 million from 1978 to 1982 (employing an exchange rate of 220 yen to the dollar). Production of the 767 components is handled by the Japan Commercial Aircraft Corporation (JCAC, composed of the same firms as the JCTDC group that developed the 767 venture). In recent months, unexpectedly slow sales of the 767 have created some financial difficulties for the consortium:[10]

CAC, which comprises Mitsubishi, Kawasaki, and Fuji Heavy Industries, had been planning on a production rate for fiscal 1983 of up to 10 aircraft a month. Due to the sales slump, the rate will be on 2-3 a month, cutting CAC's profits and reducing its ability to repay the government, the MITI official said.
   The reduced production rate also has left the CAC companies with excess capacity — a particularly heavy burden in Japan, where firms traditionally do not lay off employees, even in hard times.

More recently, Japanese firms have become involved in a multinational project for the development of a fuel-efficient jet engine. Japan Aero Engines, a consortium made up of Ishikawajima-Harima, Mitsubishi, and Kawasaki Heavy Industries, was established with MITI support in 1971 to explore the development of a 20,000 pound thrust high-bypass engine. Testing activities in England by the consortium led to the involvement of Rolls-Royce as a co-equal partner in the development programmes for what was then known as the RJ500. Mushrooming development costs and increasingly fierce competition in this segment of the engine market (expected to provide the engines for a 150-seat aircraft) contributed to the decision by Pratt and Whitney, along with MTU of Germany and Fiat of Italy, to join with the existing

RJ500 consortium in the development of a slightly larger engine (now known as the V2500). Japan Aero Engines holds an equity share of 19.9% in the consortium, and is to be responsible for 23% of the work, focusing primarily on the engine's compressor and fan. MITI financial assistance to this consortium has again been channelled through *hojokin*, totalling roughly $62 million, slightly more than 50% of the development costs during the period 1980-82.

Concurrently, the successor to the JCTDC, the Japanese Aircraft Development Company (made up of the same three firms, with additional financial participation by Nissan) has been conducting design work on a 150-seat aircraft, the so-called YXX. Negotiations have been going on with Airbus and Boeing over participation as a partner, rather than a subcontractor, in all phases of the production and sales of the new aircraft. As of this writing, it is likely that the JADC will form a partnership with Boeing, in which the Japanese group will have an equity share of roughly 25%. It should be stressed that Japanese participation in this project would be qualitatively different from the 767 venture, in which Japanese firms were involved to only a modest extent in design activities, focusing instead on production of components. In the YXX (or 7-7, in Boeing parlance) project, the Japanese consortium is expected to participate in all phases from design through sales and product support. While an agreement is likely to be announced soon,[11] the actual project may be delayed by the severe uncertainties that characterize the current market outlook for a 150-seat aircraft.

The current structure of the Japanese commercial aircraft industry
An aspect of the Japanese commercial aircraft industry that is central to an understanding of industry structure or government policy is the small size of the Japanese domestic market. Within the United States, Canada, the UK, Italy, France, West Germany and Japan, as of 31 December 1981, there were roughly 2,700 commercial passenger aircraft operated by the major carriers (e.g. the seven trunk carriers in the USA). Of these 2,700, Japanese scheduled carriers accounted for approximately 210, or less than 10% (even Japan Airlines, operator of fifty 747s, the largest single such fleet, accounts for less than 10% of the total number of 747s produced). The domestic market for other types of aircraft within Japan is even more underdeveloped, relative to the population and income levels of the country. General aviation and business aircraft utilization is hampered by the virtual absence of an airport infrastructure. In sharp contrast to the USA, where federal and state funds have supported the construction of navigation aids and

airfield facilities for smaller aircraft, Japan as of 1981 had only ten airports for the exclusive use of general aviation aircraft. The underdeveloped state of general aviation in Japan is due in large part to the excellent high-speed passenger rail network in the country, which is highly competitive with scheduled air transportation, especially in the populous Tokyo-Kyoto corridor.

The small size of the domestic market is reflected in the small size of the aircraft industry overall. Sales of both military and commercial aircraft amounted to $1 billion in 1981, or roughly 4% of total sales of the US aircraft industry. The industry is highly concentrated. The major firms are Mitsubishi Heavy Industries (which accounted for 49% of the total 1981 value of shipments), Kawasaki Heavy Industries (21%), and Ishikawajima-Harima Heavy Industries (21%).

The aircraft activities of these firms also are dominated by military sales to a much greater extent than is true of the US industry. Military sales during the late 1970s and 1980s generally have accounted for at least 80% of total industry sales (versus roughly 50% in the US industry). Such heavy dependence on the Japanese military market, while highly profitable to the firms thus engaged, imposes severe limitations on the growth of total aircraft demand. These limits stem from the informal but thoroughly established ban on Japanese weapons exports (which would not in any event be feasible for co-production aircraft), as well as the equally informal and binding ceiling of 1% for the share of Gross National Product that may be employed for total military expenditures. Any growth in the aircraft market must perforce come from an expansion of the commercial aircraft market.

Aircraft sales represent a small share of the total sales of the major Japanese firms, by comparison with the major US manufacturers. Total aerospace sales in 1981 were less than 10% of the total sales of each of the four major Japanese aircraft firms. Within the US industry, General Electric is the only firm that approaches a comparable level of diversification, with 1983 engine sales accounting for roughly 11% of total revenue. In contrast to major US aircraft firms, then, the higher level of product diversification within the Japanese aircraft producers provides some protection against the potentially disastrous consequences of commercial failure in the aircraft market.

Firm-financed R&D expenditures and the role of subcontractors are two aspects of the Japanese aircraft industry which merit particular attention. In contrast to the situation within the US industry, subcontracting appears to play a substantially greater role in military, as compared with civilian, aircraft production. One estimate[12] places the proportion of aircraft value which is subcontracted at roughly 60% for Japanese military aircraft. Moreover, much of the subcontracting of military aircraft components and assemblies is carried out by the large aircraft firms, rather than by much smaller firms, as is the case in other Japanese industries. Subcontracting in commercial aircraft production appears to adhere more closely to the conventional pattern of reliance on smaller firms. These contrasting patterns of subcontracting in civilian and military aircraft production may reflect efforts by the Japan Defense Agency to support several military contractors by distributing military procurement contracts among them. Similar patterns of distributed government purchases have, of course, been observed in the United States.[13] The intense interfirm competition that provides other sections of Japanese manufacturing thus may be offset somewhat (although by no means completely) by the military procurement strategy of the Defense Agency.

The major Japanese aircraft firms spend substantially less on R&D, relative to their size, than do US aircraft producers. Own-financed R&D amounts to 8-10% of the sales of the major US firms, but is generally less than 2% of the sales of Japanese producers.[14] Low levels of firm-financed aircraft research within Japan are by no means offset by high levels of government research expenditure, as is discussed below. Indeed, a key weakness of the Japanese aircraft industry at present is the almost complete absence of large-scale sophisticated test apparatus and facilities.[15]

Government policy and industrial development in aircraft: Japan
One of the most interesting issues in any analysis of the Japanese aircraft industry is the role of government policy. As we argue below, Japanese government policy towards the commercial aircraft industry displays some unusual areas of contrast and similarity with industrial policy in other sectors of the Japanese economy. The commercial aircraft industry is not easily accommodated within the "catch-up" (or "infant industry") strategies which have been employed in such other export industries as steel or automobiles. While Japanese government policy displays an awareness of this fact, the policy remains incomplete and inconsistent. In addition, the contrast between Japanese and US government polices towards their respective national aircraft industries raises some issues of interest.

We begin by summarizing the elements of the "catch-up"

industrial strategy. This discussion is followed by a brief analysis of the basis for MITI's recenty expressed interest in commercial aircraft as a desirable industry for future Japanese participation. Finally, the character of MITI policy with the aircraft industry is considered, in order to contrast and compare it with policies towards other Japanese industries.

The general characteristics of the catch-up or infant industry policy framework applied by MITI and the Ministry of Finance to such industries as steel, computers, and microelectronics include the protection, through tariffs or administrative suasion, of the domestic market in the industry's infancy. The importance of the Japanese domestic market as a launch base for export industries has been stated forcefully by Myohei Shinohara, former Director of the Economics Research Institute within the Economics Planning Agency:[16]

One of the basic factors which made export promotion easier in Japan was the huge domestic market of approximately 100 million people. If the domestic market expands in line with or ahead of export expansion, a product with a higher rate of expansion would be subject to a considerable reduction in unit cost through mass production, thus allowing an increase in exports. In other words, even though the relationship between the expansion of the domestic and export markets might have been that of a trade-off on an extremely short-term basis, it proved to be highly complementary for the mid- and long-term. The existence of a feedback relationship between expansion of domestic demand and exports resulted in high growth in Japan.

Another element of this policy framework, one in which MITI has been heavily involved, is the technical development of industry. During the 1950s and early 1960s, as Johnson[17] and others have noted, much of MITI's power derived from its control (shared with the Ministry of Finance) over the allocation of foreign exchange in the period of limited convertibility for the yen.[18] During this period, MITI played a major role in the identification of foreign industrial technologies for import into Japan, the negotiation of favourable terms for the licensing of the technologies from foreign patentholders, and, of central importance, the liberal licensing of these foreign technologies within Japan.[19] The goal was the provision of a pool of technological knowledge which would be relatively accessible to firms throughout a given industry.

In the face of declining direct and indirect powers of control over foreign exchange and investment funds, MITI more recently has encouraged joint research ventures among firms within a given industry (notably, computers and large-scale integrated circuits, as Okimoto has noted).[20] To some extent, the increased emphasis on domestic sources of industrial technology reflects growing scientific and technological strength. The goal of this new policy remains the same, however — the nurturing of an industry-wide knowledge base that will support vigorous interfirm competition in product development and manufacture.[21] As Peck and Goto have pointed out, interfirm competition in many export-oriented sectors of Japanese industry is quite intense:[22]

While there is a stereotype of Japan as the home of cartels and government limitations on competition, the extent of competition appears significantly high for the industrialized oligopolistic economy that is found in all major market economies.

The pattern of prices in the high technology industries supports this view. Wholesale prices in such industries as iron and steel, transportation equipment, electric equipment, chemicals and petroleum products declined through the sixties, even though wages were rising at an annual rate of more than 10 per cent.

This combination of interfirm cooperation in the development of industrial technology and interfirm competition in product design, sales and marketing has proven extremely potent.

More recent initiatives in the area of industrial policy have emphasized the need for a stronger basic research foundation to support the development of indigenous Japanese technological strengths in the areas of biotechnology, computers, and telecommunications. As the Economic Planning Agency has noted, this goal of technological creativity reflects the fact that in many industrial applications, Japanese technology is the equal or superior to that of other industrialized nations:[23]

there is no longer any room for the "catching-up type" of thinking which seeks models of industrial structure, technological innovation and national life in Western nations, and positively introduces them into Japan to promote the progress of the Nation ... From now on Japan must open up a path of its own both economically and socially.

The specific details of this new industrial strategy are not well developed, but it is possible that the role of MITI or the Ministry of Finance will diminish somewhat. Indeed, the future path of Japanese economic development is likely to be one in which few of the institutions which have been of importance in the past 25 years will retain their significance. Trading companies, bank lending (as opposed to equity finance), and "administrative guidance" all seem likely to recede somewhat in importance.

The commercial aircraft industry occupies a peculiar position within the current framework of Japanese industrial policy While the general strategy being pursued by the major aircraft firms and MITI is quite consistent with the "catch-up" model, in other respects that model is inappropriate. Due in part to the nature of the production technologies and know-how that are necessary to become a major participant in the commercial aircraft industry, greater reliance on the "basic research" model might be more appropriate. However, current policy is not following this path.

In its *Vision of MITI policies in the 1980s*, MITI identified commercial large transports as a key industry for the future development of the Japanese economy:[24]

The aircraft industry is a typical knowledge-intensive industry, characterized by high added value and far-reaching technological spin-off. It will play an important role in the national plan to remold Japan's industrial structure into an innovative knowledge-intensive type.

At present the aircraft industry is smaller in scale in Japan than in advanced Western countries and relies excessively on demands for defense industry. It should direct more attention to the manufacture of planes for civil transportation which has a big future.

It seems realistic that the private sector should hear the ultimate risks involved in an aircraft development product, but for the time being the government will subsidize products on the condition that a percentage of profits be contributed to the government, contingent on success.

It is hoped that Japan will build up a system for basic research and development of aircraft engineering so that it may be fully ready for the expected technological innovation in the 1990s for the manufacture of the next generation aircraft. Development of aircraft engineering must be conducted on the initative and assistance of the government as it involves highly sophisticated and complex technology.

The attributes of this industry's output which rendered it attractive for future development, according to MITI, included high value-added characteristics, high world income elasticity of demand, minimal environmental pollution, and high knowledge intensity. Two other characteristics of the industry cited by MITI deserve particular attention. The first concerns the links between aircraft and other high technology industries. As was noted above, these linkages are numerous and significant, and constitute a key dimension of this industry in the view of MITI policymakers.[25] In addition, the technology of commercial aircraft was described as one in which a Japanese comparative advantage would not be undercut by such Asian competitors as South Korea or Taiwan as rapidly as had been true of other industries (e.g. steel). Obviously, this argument is a two-edged sword — the prospective difficulty faced by South Korea or Taiwan in acquiring aircraft production and design skills will also hamper Japanese efforts to acquire such skills from US or European producers.

Despite this public identification of aircraft as a strategic future industry (a statement largely responsible for the flurry of US commentary noted above), Japanese industrial policymakers and firms have, of course, been involved in the aircraft industry for some time. Both the YS-11 and the extensive military co-production programmes within the aircraft industry were classic "catch-up" strategies. Both policies were efforts to develop aircraft design and production expertise by producing for the domestic market. However, the YS-11 experience clearly indicated that this domestic market is far too small to support by itself the introduction of a new aircraft. The small domestic aircraft market is a crucial element distinguishing aircraft from such other successful Japanese exports as automobiles or steel. The input of the co-production of military aircraft also has been limited by the small domestic market for military aircraft.

While the co-production of aircraft by Japanese firms appears to have improved production technologies within the participant firms, as well as supporting the emergence of a small group of vendor firms, the impact of this programme on the design capabilities of Japanese aircraft firms has been modest. In contrast to commodities such as steel or even consumer electronics, the key component and source of value added (and thus profit) in commercial aircraft is in the design of aircraft, a task requiring high levels of fundamental knowledge and research.[26]

Inasmuch as the infant industry strategy for encouraging the development of the aircraft industry appears to have severe deficiencies, the recent 767, YXX and V2500 projects have been instituted as alternative means of development. The goals, revealed most clearly in the YXX programme are the acquisition of expertise in all phases of aircraft design, manufacture, and sales.[27] It is also true, of course, that a fuel-efficient 150-seat transport would serve as an excellent replacement for the ageing YS-11 fleet that still serves the Japanese domestic market.

Several aspects of current Japanese policy in the aircraft industry are of interest. MITI policymakers are interested primarily in entry by Japanese firms into the design and production of large commercial transports, rather than general aviation or commuter transport aircraft. This goal is pursued despite the demonstrated success of at least one Japanese firm in the general aviation and business aircraft market (Mitsubishi Heavy Industries, maker of the MU-2 and the Diamond 300). The market outlook in commuter aircraft also is rather more robust than that for large commercial transports, reflecting the impact of deregulation of air transportation in the United States, and increasing demand from industrializing nations for small, short-haul aircraft. Indeed, as was noted above, the financial rewards thus far for the Japanese participants in the 767 venture have been very modest. Nonetheless, MITI policymakers have suggested that technological supremacy is less central to the sales of these other aircraft types, implying lower per-unit profitability. General aviation and commuter aircraft design also demands a lower level of technological expertise, meaning (among other things) that a Japanese technological lead in this industry segment is likely to be shorter in duration.[28]

The desire of the Japanese CADC consortium to be involved in all aspects of the development of the YXX also means forgoing the efficiencies and resource savings associated with specialization in a single area of aircraft design or construction. If the long-run goal of policymakers in this area is the establishment of Japanese aircraft firms as participants in international consortia, specialization is likely to be far more effective. For example, the sophisticated instrument panel displays of the A310 were developed and produced in Japan. Such national specialization has emerged in the development of the European Airbus. The apparent decision to avoid specialization points to a conflict between the use of international consortia as mechanisms for industrial development and technology transfer, and the commercial success or failure of such consortia. This theme is discussed below.

The current policy framework also places little or no emphasis on the funding or performance of basic aeronautical research. Despite the recent outpouring of statements from Japanese officials concerning the need for more "creative" fundamental research, little or none now is performed within the Japanese aircraft industry. MITI is providing loans to cover up to 75% of the costs of preliminary design and development work on the V2500 engine or the YXX transport. However, public funding for non-mission-oriented aeronautics R&D within Japan is very modest. Neither the National Aeronautics Laboratory nor the Japanese Defence Agency is a significant source of research funding, and the number and sophistication of engine and airframe test facilities within Japan is low, as was noted above. MITI policy towards aircraft thus displays some similarities to the earlier "catch-up" strategy, in focusing on the acquisition of existing product and production technologies. Whether aircraft are well suited to such a strategy is unclear.

Finally, MITI policy towards the aircraft industry is unusual in supporting low levels of interfirm competition and substantial interfirm cooperation. The allocation of shares of work among consortium members is determined through negotiation rather than competition. Rather than encouraging interfirm cooperation in research and development, while supporting competitive product design and manufacture, the aircraft industry consortia cover design, development and production. This essentially non-competitive policy seems to reflect the influence of several factors. As was noted above, development costs in aircraft are sufficiently large that it may be exceedingly difficult for a single firm to sustain them.[29] In addition, it is possible that the environment of reciprocity in the allocation of military aircraft contracts and subcontracts among these firms may have encouraged simlarly noncompetitive behaviour in commercial aircraft development and production. Whatever the reason, the lack of interfirm competition in the production phase of this industrial development strategy distinguishes it from the pattern observed in other Japanese export industries.

The commercial aircraft industry may represent a transitional case of Japanese industrial policy. As was noted above, the structure of MITI policy for the industry is similar in many respects (though not all) to previous "catch-up" strategies in other industries. However, both the world market and the technology associated with commercial aircraft are such that the "catch-up" strategy is likely to be, and has been, a mixed success at best. The resulting policy of reliance on international joint ventures thus appears to be something of a compromise between a "catch-up" strategy and a more long-term policy aimed at the strengthening of indigenous technological resources. Neither the goals nor the policies of MITI in this industry appear to be entirely internally consistent.

US Government policy towards the commercial aircraft industry, 1945-83: a summary and comparison with Japan
In view of the pervasive impact of government policy on the US commercial aircraft industry, a brief summary of the structure of this policy and a comparison with the Japanese case seems useful. US government policy, especially prior to 1978, had the effect of enhancing both the supply of and the demand for innovations in aircraft components and design. As is discussed below, this policy framework reflected the links between military and commercial aircraft design and manufacture. However, government policy also facilitated the development of an industry-wide knowledge base, accessible to all firms within the US aircraft industry. In its support of final product demand, as well as its encouragement of the development of a widely shared knowledge base, US government

policy prior to 1978 resembled that observed in the development of earlier Japanese export industries.

During much of the post-war period, the existence of substantial commonality between military and commercial aircraft technologies meant that federal military procurement and sponsorship of research, both of which were carried out on a very large scale, generated significant military-civilian technological spillovers, expanding and enhancing the pool of technical knowledge available for application in commercial aircraft designs. Examples of such spillovers include the high-bypass jet engine, the original turbojet engine, and the Boeing 707.

An additional and important aspect of military aircraft research and procurement was the fact that much of the resulting technological knowledge was widely diffused within the industry. This widely diffused knowledge base stemmed from the fact that federal regulations required liberal licensing and/or the assignment of patent rights to the government, as well as the fact that firms frequently formed temporary consortia in competing for military contracts. Inter-firm cooperation in R&D and the development of a widely shared knowledge base, both of which are characteristic of MITI policy in a number of Japanese industries, were thus facilitated by the military research and procurement system during much of the post-war period.

Federal policy went beyond its influence on military aircraft technology. The National Advisory Committee for Aeronautics (NACA) and its successor, the National Aeronautics and Space Administration (NASA) both carried out extensive programmes of research in civil aircraft. Indeed, the aeronautics research budget for NASA in fiscal 1982, roughly $450-500 million, is substantially greater than the sum of research and development funds provided by MITI and the National Aeronautics Laboratory within Japan. Roughly one-half of these funds are channelled through private firms. A substantial network of publicly funded, large-scale test facilities, available for use by firms for a subsidized charge, were constructed and are operated by NASA. Similarly to military research, NASA-funded research data and results were widely available within the industry. In addition, NASA projects not infrequently have involved two or more erstwhile competitor firms, encouraging, to at least a limited degree, the pooling of research efforts and results of firms that were vigorous competitors in the production and sales of airframes or engines. An additional fact or working in favour of a widely shared technology base within the US aircraft industry was the sponsorship by NACA of a liberal system of cross-licensing of patents. This system was disbanded in 1975, due to the objections of the Antitrust Division of the Justice Department.[30]

The regulatory regime of the Civil Aeronautics Board, which was substantially weakened in 1978, supported a strong market demand for innovation in commercial aircraft. CAB regulation, which blocked entry into the long-haul market and largely prohibited price competition between carriers, led to competition among carriers in the area of service quality. This competitive environment in turn encouraged airlines to acquire new generations of aircraft — an early delivery position in the order queue was deemed highly desirable, and the "launching" of new aircraft designs and models was encouraged. This environment encouraged rapid rates of technical change and product development in commercial aircraft. Moreover, sustained dominance of the US aircraft industry by a single firm was largely avoided prior to the 1970s.

The changes in the policy environment of the commercial aircraft industry occurring during and after the 1970s are likely to lead to lower rates of new product development, and have encouraged a substantial decrease in the number of producers. The relationship between military and civil aviation aircraft technologies has changed somewhat. Partly due to the absence of major military initiatives in the large transport area, technological spillover from military to commercial applications has been reduced. Commercial aircraft no longer benefit from military research and procurement to the same extent as was previously true. On the civilian technology side, NASA aeronautics R&D budgets have declined in real terms for much of the past 15 years — the recently announced space station programme is likely to continue this trend. The dissolution of the patent pool in 1975 also may have contributed somewhat to a reduction in the size and accessibility of a common industry knowledge base. Finally, the demise of CAB regulations, opening the industry to entry and competitive pricing, has changed the terms of competition among airlines. With a decline in the importance of service quality, relative to price, as a means of attracting customers, the demand for new aircraft models has been reduced. For aircraft producers the assembly of a customer base for the initial launch order for a new aircraft now is more difficult.

The postwar commercial aircraft industry in the United States was a beneficiary of a policy framework that was in many ways similar to that in various Japanese industries in this period. Public support was provided for both the supply of technological knowledge and the demand for its embodiment in new products.

The analogous policies in the Japanese context are the management of technology imports and the protection of the domestic market. Of equal importance, perhaps, is the achievement of a balance of sorts in the US aircraft industry between interfirm cooperation in fundamental research and interfirm competition in product development and sales.

The performance of the aircraft industry in the post-war US economy suggests that these general principles of industrial policy are no less applicable in the US than in Japan. However, several caveats should be mentioned. This policy framework in the United States was not adopted as a purposive, coherent package of measures aimed at a specific industry. The US political environment, in which distributional issues play a central role in policy formulation, is inhospitable to the explicit formulation of industry-specific strategies.[31] Moreover, the direct and indirect costs of this policy structure were substantial. Military procurement and research, as well as the NACA/NASA research programme, have consumed tens of billions of public funds over the past three decades. The CAB regulatory framework also imposed substantial indirect costs on consumers, in the form of higher prices for air travel. The welfare impact of this policy framework for aeronautics thus seems ambiguous at best.

The outlook for the Japanese aircraft industry
The aircraft industry occupies a peculiar status within Japanese manufacturing industries, in that a significant technology gap between Japanese and other producers coexists with an insufficient domestic market to support a "catch-up" industrial strategy. In contrast to the symmetrical policy which prevailed through 1977 in the United States, one affecting both the demand for and the supply of aircraft technology, MITI's general strategy appears to be one of limited support for a "technology push", with virtually no demand pull. Moreover, the support for technological development in aircraft that is offered by MITI and other government sources is modest by comparison with the USA. In 1979, total government R&D expenditures on both civil aviation and space exploration in Japan amounted to roughly $262 million in 1975 dollars, while R&D expenditures by the US government on civil aviation alone equalled $344 million in 1975 dollars.[32] Clearly, Japanese subsidy of aircraft R&D is substantially below the level observed in the USA. In view of the fact that the privately financed research expenditures of Japanese aircraft firms are below those of the US firms, it is difficult to see how the gap in aircraft design technology

in particular can be closed rapidly, if at all. It is true that the focus of much Japanese public R&D expenditure in this area on the development of specific aircraft or engines raises the possibility that Japanese research expenditures complement more fundamental reserach (e.g. that performed and widely disseminated by NASA). However, the greater mission orientation of Japanese R&D priorities is likely to be detrimental to the longer-term development of technological capabilities in the aircraft industry.

The focused nature of the current research and development efforts in the Japanese aircraft industry reflects the fact that, to a great extent, commercial strategy within this market is being overseen closely by MITI. This close oversight (not to say direction) is another factor responsible for the absence of interfirm competition in product development and sales. Rather than the relevant firms making the competitive judgements individually, MITI, in consultation with these firms, is determining product development strategy. Intervention at a point in the product development process that is so close to commercialization is rare for MITI, and more rarely successful. An analogous policy was the ill-fated attempt to bring about a merger of the major Japanese auto producers to manufacture a "people's car".[33] The MITI strategy in aircraft is both risky and unusual. In choosing to gamble heavily on the commercial prospects of the 150-seat plane and a new high-bypass engine, the Japanese policy faces a substantial chance of commercial failure. A more competitive, less closely directed product development strategy exploring other market segments and relying on a hgher level of non-mission-oriented research funding, might be more prudent.[34]

International joint ventures in new product development
The use of international consortia as mechanisms for product development, manufacture and sales, as well as technology transfer, raises a number of interesting issues. The first concerns feasibility. While there exist examples of successful international joint ventures in product development and manufacture (within aircraft, the GE-SNECMA and Airbus enterprises are cases in point), there are equally numerous examples of ruinous cost overruns, design disagreements, and commercial failures. (Lorell[36] provides a number of case studies of European joint ventures in the military aircraft sector; the Concorde and McDonnell Douglas-Fokker fiascos are other examples.) Tasks of management, cost control and design that are at best challenging in a single firm can become overwhelming in a multinational consortium. In this regard, the

prospective YXX venture, whatever its commercial prospects, is likely to prove more feasible to manage and organize, inasmuch as it involves an extension of an already established working relationship between clearly identified junior and senior partners (JCADC and Boeing, respectively). The V2500, involving several firms of comparable technological sophistication, may prove more difficult. However, both of these joint ventures are also working with relatively stable technologies and desired characteristics, rather than pushing the outer reaches of the state of the art and attempting to accommodate a wide array of conflicting preferences (as in the Tornado combat aircraft). This difference suggests some grounds for optimism.

However, the amount and effectiveness of technology transfer through such consortia raises very thorny issues of conflicting goals and incentives. The most obvious conflict arises between the desire of technological leaders to retain, and other consortium members to obtain, advanced technological knowledge. In the established CFM joint venture in aircraft engines between General Electric and SNECMA, for example, the most advanced section of the engine, the so-called "hot section", has always been produced solely by GE — no sharing of the technology of this key component between the venture partners has taken place. Similarly, in the prospective V2500 consortium, the "hot section" is being produced by Pratt and Whitney, with no expectation of sharing this technology. The extent of the technology transfer that will be realized through such a joint venture is very uncertain.

Technology transfer questions were raised by both Pratt and Whitney and Rolls-Royce early in the consortium negotiations, according to Samuel L. Higginbottom, chairman and president of Rolls-Royce Inc. Both companies feel there will be a minimum of exchange of proprietary data in the final assembly process.

"We will have to know the interfaces, and there obviously will be some exchange of data involved in that," Higginbottom said. "But we will not have to get into the details of technology..."

"It took a lot of work to match the technology split to the work-sharing formula," one official said. "But now that it is completed, final assembly essentially will involve bolting together the separate modules."[35]

The V2500 consortium as presently constituted does appear to adhere to a well defined division of labour, with participant firms specializing in various components of the engine technology. As noted above, such an organizational structure is likely to reduce technology transfer. This type of structure, while more manageable, also appears rather different from the proposed YXX, in which

Japanese firms hope to be involved in all phases of design, promotion, and sales. The management of such a non-specialized form of participation can prove difficult. Indeed, conflicts over technological transfer may also become more intense, by virtue of the less precise delineation of responsibilities.

The international joint venture mechanism thus carries some serious flaws as a vehicle for industrial and technological development of the Japanese aircraft industry. In essence, factors making for greater managerial and commercial feasibility work at cross-purposes with the factors enhancing the extent of technology transfer and/or learning. While stable technological boundaries and a clearly delineated division of labour work to enhance the feasibility of joint ventures, they may also undercut technology transfer. Where the key technologies are materials and production processes, such technology transfer is likely to happen in any event. Where (as in aircraft) the key factors are design and other forms of research-based expertise, transfer through a multifirm consortium may be less effective in such a context.

Issues for policy and theory

The Japanese commercial aircraft industry is unlikely to become a major factor in the international commercial air industry in the near future. However, the competitive threat to the US aircraft industry posed by Japan is unlikely to affect the major producers. The increasingly international character of subcontracting and the components trade suggests that the smaller component and supplier firms within the US industry may face a more competitive environment in the near future. However, the international character of the markets faced by these firms (revealed in the high US content of aircraft such as the Airbus or the Embraer Bandeirante, produced respectively by Europe and Brazil) means that protectionist measures will benefit no one, least of all these US supplier firms. The contribution of this segment of the industry to the national defence industrial base and/or a mobilization surge capacity may provide an argument for some limited system of preferential defence procurements. However, the notion of industrial mobilization for a lengthy period of conventional war in the nuclear age seems far fetched, to put it mildly.

The increasingly international character of research and product development efforts in the aircraft industry raises other issues concerning the conceptual framework for the analysis of multinational firms and investment. The celebrated product cycle model[37] has been employed widely to analyse the basis for the existence of multinational firms and the associated flows of direct foreign investment. In this model, differences in the consumer demand profile of various national markets gave rise to differences in the abilities and attitudes of the national firms serving those markets. Such differences in firm-specific attributes in turn led to direct foreign investment as the most effective means of exploiting such firm-specific assets (which by their nature are not easily traded among firms).[38] Within this model, which was applied primarily to consumer goods, firms are essentially passive channels for the expression of differences among the various national markets of the world economy. In treating the firms as tightly constrained black boxes, the product cycle theory demonstrates a close link with the neo-classical economic view of the firm.

However, the recent international activities of aircraft, computer and (a somewhat special case, due to quotas on the import of Japanese autos) automobile firms display some points of significant disagreement with the predictions of the product cycle model. International flows of capital *per se* are rather less significant in these ventures, while the establishment of wholly owned direct foreign production facilities is even rarer. Instead, one finds a profusion of jointly owned or controlled ventures, many of which do not have as their central focus the manufacture, but rather the development, of new products.

Rather than capitalflows constituting the key feature of these new forms of multinational activity, technological information and data, design capabilities and managerial expertise appear to be the central components. In addition, the specific characteristics of the home markets of the various participant firms have little to do with the firm-specific attributes being exploited through such agreements. In effect, the firm-specific attributes of the participants no longer are "country bound" to the same extent as previously. Rather than passive actors, it is more appropriate to view these firms as actively seeking to develop firm-specific, non-transferable attributes as a means to the end of probability. We do not intend to suggest that firm behaviour has changed, but rather than changes in the international economic environment have rendered firm-specific attributes that are not based on national markets alone much more significant.

These changes have affected both the supply of factors and the nature of demand. On the supply side, particularly in a comparison of the industrialized nations, there are growing similarities in the supply and relative prices of physical and human capital. Simultaneously, the profile of product demand in the various national components of the world market is becoming more honogenous, and less dominated by any single market (e.g. the declining US share of world demand for transports). Both of these influences have made much more attractive the pooling of firm-specific attributes in joint ventures. Inasmuch as these trends bid fair to continue, we may expect to see a growing number of such arrangements in a wider range of industries. An intriguing question is the extent to which these joint ventures will serve as devices for the transfer to a variety of (as yet) firm-specific capabilities. This appears to be an important motive underlying Japanese participation in aircraft joint ventures, but it is a motive unlikely to be shared by all of the participants.

Notes

1. US General Accounting Office, *US Military Coproduction Agreements Assist Japan in Developing its Civil Aircraft Industry* (Washington DC, US Government Printing Office, 1982), p.17. See as well the report to the Subcommittee on Trade of the House of Representatives' Ways and Means Committee, which argues that "The Japanese aerospace industry appears to be preparing for takeoff. MITI is the pilot, urging and cajoling in order to raise the domestic industry from a technologically backward assembler industry of US-designed military hardware to a full fledged independent maker of commercial jet transportation ... Once Japan has her own commercial aircraft to sell ... sales to Japan will probably be cut drastically and US producers will face another major competitor. US House of Representatives, Subcommittee on Trade, Ways and Means Committee, *United States-Japan Trade Report* (Washington DC, US Government Printing Office, 1980), pp.44-45.

2. See J.W. Kendrick, *Productivity Trends in the United States* (Princeton University Press, 1961); J.W. Kendrick, *Postwar Productivity Trends in the United States 1948-73* (New York, Columbia University Press, 1973); or B.M. Fraumeni and D.W. Jorgenson, "The role of capital in US economic growth 1948-76", in G.M. von Furstenberg (ed.), *Capital, Efficiency and Growth* (Cambridge, Ballinger, 1980).

3. All figures taken from Aerospace Industries Association, *Aviation Facts and Figures 1983/84* (New York, McGraw-Hill, 1983).

4. The 1981 study of aeronautics R&D by the Office of Science and Technology Policy, *Aeronautical Research and Technology Policy: Final Report*, vol.2 (Washington DC, Executive Office of the President, 1982) concluded that "the aeronautics industry is characterized by high research intensity and a wide technology base. That is, aeronautics depends on R&T [research and technology] performed within the aeronautics industry and on R&T performed by virtually every other high technology industry". Statistical support for this characterization may be found in a recent study by the International Trade Administration of the US Department of Commerce, analysing "embodied" research intensity in various American industries. Embodied research intensity simply attempts to account for R&D expenditures that are incorporated in the inputs purchased by a given industry. The aircraft industry ranked third among US manufacturing industries on this index with an embodied research intensity

of 15.4% of the value of shipments in 1980, exceeded only by missiles and spacecraft and electronic components. *An Assessment of US Competitiveness in High Technology Industries* (Washington DC, US Government Printing Office, 1983), p.42.

5. This paragraph draws on D.C. Mowery and N. Rosenberg, "The Commercial Aircraft Industry", in R.R. Nelson (ed.), *Government and Technical Progress: A Cross-Industrial Analysis* (New York, Pergamon, 1982).

6. R. Miller and D. Sawers, *The Technical Development of Modern Aviation* (London, Routledge & Kegan Paul, 1968).

7. See, among other estimates, that in *Business Week*, "A jet engine creates an unlikely alliance", 13 September 1982, p.47.

8. Recent instances of interfirm cooperation in new aircraft and engine development include the General Electric-SNECMA consortium and the seven-firm, five-nation V2500 project in engines. The European Airbus, as well as the prospective joint venture involving Boeing and three Japanese firms in the development of a 150-seat transport, are examples of the same phenomenon in airframes. Subcontracting also has expanded greatly. According to J. Rae, *Climb to Greatness* (Cambridge, MIT Press, 1968), subcontracting of airframe assembly in the 1920s "constituted less than 10 per cent of the industry's operations" (p.83). The production of the Lockheed Electra (in the 1950s) subcontracted 30-40%. Production of the Boeing 747 entailed subcontracting 70% of the assembly of the aircraft; this percentage is higher for the 767, where subcontracting also was extended somewhat into the design and development phase, and involved greater financial participation and assumption of risk.

9. R.L. Eckelmann and L.A. Davis, *Japanese Industrial Policies and the Development of High Technology Industries: Computers and Aircraft*, prepared for the Office of Trade and Investment Analysis, International Trade Administration, US Department of Commerce (Washington DC, US Government Printing Office, 1983).

10. "Japan setting higher aerospace goals", *Aviation Week and Space Technology*, 21 November 1983.

11. According to *Aviation Week and Space Technology*, the announcement and signature of a memorandum of understanding is expected "within weeks", 23 January 1984, p.30.

12. Interview with Atsuhiko Bansho, Managing Director, Society of Japanese Aerospace Companies, June 1983.

13. The list of examples is a lengthy one, but includes the purchase from McDonell Douglas of modified DC-10s for use as tankers and, arguably, the F-111 award to General Dynamics in 1963. S.L. Carroll asserts that government procurement "provides the safety net that catches a plummeting airframe company. Large backlogs of government contracts furnish rather steady income during periods when commercial activities make sales and earnings volatile ... the government simply will not allow a major defense contractor to fail completely, whatever its commercial sins." "The market for commercial aircraft", in R.E. Caves and M.J. Roberts (eds), *Regulating the Product* (Cambridge, Ballinger, 1975), p.162.

14. Interview, Ministry of International Trade and Industry, June 1983.

15. The RJ500 joint engine venture originated because of the need for the Japanese consortiums to utilize the test facilities at Rolls-Royce in England — no comparable installation existed within Japan. Interview, Ishikawajima-Harima Heavy Industries, June 1983.

16. M. Shinohara, *Industrial Growth, Trade, and Dynamic Patterns in the*

Japanese Economy (University of Tokyo Press, 1982), pp.22-23.

17. C. Johnson, *MITI and the Japanese Miracle* (Stanford University Press, 1982).

18. In the aftermath of the restoration of convertibility, the still restricted nature of the Japanese capital market has continued to confer considerable powers of moral suasion upon the Ministry of Finance and MITI in their dealings with the major Japanese banks, allowing for some influence over the allocation of investment.

19. L.H. Lynn, *How Japan Innovates: A Comparison with the US in the Case of Oxygen Steelmaking* (Boulder, Westview, 1982), provides a detailed account of MITI's involvement in the acquisition by Japanese steel firms of basic oxygen furnace technology.

20. D.T. Okimoto, *Pioneer and Pursuer: The Role of the State in the Evolution of the Japanese and American Semiconductor Industries*, occasional paper, Northeast Asia-United States Forum on International Policy, Stanford University, 1983.

21. It has been hypothesized by Okimoto and others that the system of lifetime employment and other obstacles to interfirm labour mobility in Japanese industry have the effect of redirecting the interfirm diffusion of knowledge and technology, thus increasing the importance of such policies as those discussed here.

22. M.J. Peck and A. Goto, "Technology and economic growth: the case of Japan", *Research Policy*, 10, 1981, pp.222-243, at p.238. It should be noted that MITI's policies have not been consistently supportive of intense interfirm competition. The attempts of MITI to force mergers among Japanese automakers suggest some considerable trepidation concerning the benefits of competition.

23. 0, trans. by *The Japan Times*, (Tokyo, 1983), p.18.

24. Ministry of International Trade and Industry, Industrial Structure Council, *The Vision of MITI Policies in the 1980s* (Tokyo, Industrial Bank of Japan, 1980), pp.291-292.

25. This general basis for the choice by MITI of a particular industry for focused assistance has been noted by Hadley in a survey of Japanese industrial policy: "Japanese target industries have been selected not only for their own importance, but for their ramifying effect on other industries." E.M. Hadley, "The Secret of Japan's Success", *Challenge*, 1983, pp.4-10.

26. This assessment is consistent with the recent growth of subcontracting by the major US aircraft producers, discussed above.

27. It is also likely that political obstacles to successful export of an all-Japanese large commercial transort would be raised by other industrialized nations. Government ownership or control of many international carriers means that new aircraft purchase decisions frequently are highly political in nature.

28. Interview with MITI personnel, June 1983.

29. Indeed, the primary reason for Nissan's participation in the YXX project is the ability of Fuji Heavy Industries to bear its assigned share of the production costs.

30. Miller and Sawers, *The Technical Development of Modern Aviation*, described the licensing system of the Manufacturers' Aircraft Association as a system "under which all aircraft manufacturers agreed to let all their competitors use their patents. No member can have a patent monopoly on any inventions which his staff can make, or even for any invention that he may license from an outside inventor. If he takes an exclusive license, the patent has to be available to the other members of the MAA; but the original licensee can claim that he

should be granted compensation by the other licensees. Manufacturers apparently believe that it is a good bargain to give up their right to a patent monopoly in return for the protection from litigation with other companies in the industry that the right to use their patents brings." (pp.255-256)

31. See C.S. Schultze, "Industrial policy: a dissent", *The Brookings Review*, 2, 1982, pp.3-12, for a critical discussion of industrial policy proposals that examine this theme in detail.

32. See A. Rapoport, "A macro comparison of civil aviation and civil space R&D projects among major industrial countries during the 1970s", unpublished paper, Division of Policy Research and Analysis, National Science Foundation, 1980, Tables 5 and 7. Rapoport's data do not include *hojokin*. However, these amounted to no more than $40 million in 1979.

33. Throughout the 1950s and 1960s, MITI was extremely concerned about the structure of the Japanese automobile industry, arguing that the industry was too fragmented to withstand the inevitable international competition of the future. A number of efforts to rationalize the auto industry's structure followed, including an attempt in 1955 to sponsor a competition among firms to design and produce (with government subsidy) a so-called "people's car". The intent was to allow one firm to dominate the auto market. However, strong industry opposition prevented the proposal from reaching the Diet. See I.C. Magaziner and T.M. Hout, *Japanese Industrial Policy*, Institute of International Affairs Policy Paper 14, University of California, Berkeley, 1980; or US Department of Commerce, Bureau of International Commerce, *Japan: The Business-Government Relationship* (Washington DC, US Government Printing Office, 1972), for further discussion.

34. The National Aeronautics Laboratory has conducted an extensive research programme in short takeoff and landing (STOL) aircraft, culminating in the development of an experimental prototype. However, the prototype has been developed largely by the public laboratory, with modest levels of interaction with the private sector. Thus far, there have been no serious efforts to develop a commercial aircraft based on the prototype.

35. "Pratt, Rolls launch new turbofan", *Aviation Week and Space Technology*, 7 November 1983, p.29.

36. M.A. Lorell, *Multinational Development of Large Aircraft: The European Experience*, RAND Report R-2586-DR&E (Santa Monica, The Rand Corporation, 1980).

37. See R. Vernon, "International investment and international trade in the product cycle", *Quarterly Journal of Economics*, 80, 1966, pp.190-207; and R.E. Caves, *Multinational Enterprise and Economic Analysis* (Cambridge University Press, 1982).

38. For additional discussion of this point, see W.M. Cohen and D. Mowery, "Firm heterogeneity and R&D investment: an agenda for research", in B. Bozeman, M. Crow and A. Link (eds), *Strategic Management of R&D: Interdisciplinary Perspectives (Lexington, D.C. Heath, 1984)*.

NATIONAL SCIENCE FOUNDATION EXPERIENCES IN STIMULATING INDUSTRIAL INNOVATION

R.M. Colton, T.M. Ryan and D. Senich

The National Science Foundation (NSF) experiences in stimulating industrial innovation have taken two forms.

First is the development of cooperative research arrangements between industry and the university. In this form, individual companies and a university are encouraged to collaborate in basic research whereby initial technical review and funding by the foundation provides the stimulus for continued industry-university cooperation; or, a group of companies and a university are encouraged to establish a basic and applied research centre with research output being of interest to both industry and academia. In this latter case the foundation (through a two-phase approach) provides the seed money over a period of five years to establish a long-term research centre. A major objective of the centre is to become self-sustaining, that is, entirely funded by industry within the five-year period.

A second form is to fund small technology-based businesses directly, on a competitive basis, for feasibility and development studies leading to commercialization of some new product, process or service. This is also conducted on a two-phase basis, whereby NSF only funds the research component.

Both of these mechanisms are described and elaborated upon in the following sections of this paper.

Industry-university cooperative research

Background

Industry-university research cooperation is clearly a national trend. This relationship represents an active partnership aimed at re-energizing the innovation and productivity of the US industrial machine that in many ways has been the model for the free world. Industry and the university are thus cooperating to meet the scientific and technical challenges posed by our domestic needs as well as international competition. Interest in industry-university cooperative research is now more intense because of the growing perception that industrial products and services are increasingly dependent upon fundamental scientific understanding.

A recent National Science Board report on university-industry research relationships provides much detail on the cooperative relationships which presently exist (see *University-Industry Research Relationships: Myths, Realities and Potentials*, National Science Foundation, February 1983). According to the report, new arrangements reflect an optimistic mood that is grounded in an awareness that the problems and opportunities in technologically based industrial production are substantially different from those of the past. Three general factors characterize this change.

First, product and process improvement and innovation in some industries have evolved to levels of complexity that demand understanding of fundamental physical and biological phenomena and thus require much higher levels of training in and use of basic science and engineering.

Further, incremental advances in narrowly focused technical areas, characteristic of much industrial development in the past, are giving way to the use of a broad range of science and engineering disciplines on complex, often ill-defined problems, or exploitation of new analytical capabilities. Hence it is becoming increasingly difficult for any one industrial laboratory to fully encompass the requisite expertise. A partial remedy for this situation is to seek out the pertinent skills wherever they may be found in the nation's universities.

And finally, the rapid expansion of the nation's R&D system over the past three decades has diffused research capabilities over a much broader range of institutions — academic and industrial — than before. Thus it is quite unlikely that any one company could hold and maintain a leading edge on technical advance in a given area as, for example, Dupont was able to do in polymer fibres during the 1930s and later.

Programmes

The long-term goal of NSF industry-university research programmes is to improve technological innovation with the objective of:

- Developing research linkages;
- Leveraging non-government scientific and financial resources;
- Stimulating fundamental industrially relevant research; and

- Implementation of federal programmes such as the Small Business Innovation Development Act of 1982 (PL 97-219) and the Stevenson-Wydler Technology Innovation Act of 1980 (PL 96-480).

Current NSF programmes designed to accomplish these objectives include: Industry-University Cooperative Research Projects, University-Industry Cooperative Research Centers, Small Business Innovation Research, and Productivity Improvement Research. The cooperative research programmes provide mechanisms intended to encourage collaboration in research by academic and industrial scientists according to their own research priorities. Small business projects provide another focus for the leveraging of human resources in that there is significant involvement by university researchers on a consulting basis (and often on a graduate training basis) with new research-based companies. Productivity improvement research deals with understanding and management of techniques to overcome barriers to improved productivity and innovation.

Industry-University Cooperative Research Projects
The Industry-University Cooperative Research (IUCR) Projects programme encourages and advances scientific knowledge which is relevant to industrial technological innovation through the means of research cooperation between university and industry researchers.

Cooperative industry-university research projects focus on fundamental questions of science and basic engineering underlying future technological advancements. The programme does not support commercial product development. Research proposals are jointly developed by scientists and engineers in universities and industrial firms. Cost sharing by industry is required and provides a mechanism for assuring that research topics have industrial management interest. Each project utilizes the unique capabilities of both participants, and synergism results in a better research project than either could do alone.

Cooperative research projects occur most frequently in the science and engineering areas which are directly in contact with industrial technology. They are both fundamental in advancing scientific frontiers and technologically important in deepening and expanding the knowledge base upon which to build new technology. The industrial problems provide univesity researchers with challenging fundamental problems, while the academic perspective encourages industry to try more fundamental approaches than industry is accustomed to use.

The distribution of IUCR funds has been about half to the science divisions and about half to the engineering divisions. This distribution had followed proposal pressure, modified by a programme goal to involve the science fields in cooperative research, as well as the engineering fields. The reason for this goal is that basic research in engineering is normally relevant to the middle term technological changes in industry, while basic research in science is normally relevant to the long-term technological changes in industry. As for the short-term changes, industry normally takes care of this itself. The university-industry research connection is therefore for the middle- to long-term research needs of industry.

Cooperative projects in the science areas investigate the phenomena underlying important technologies, and in the engineering areas they investigate the basic principles in designing industrial devices, processes and materials. In both areas new technological feasibility and scientific instrumentation are often created.

The industries which normally participate in cooperative research are high-technology industries, since most firms in these sectors have sophisticated research capabilities and derive new technology directly from basic research. Each sector uses different fields of science and engineering. On the other hand, each sector can make contributions to different areas of science by participating in cooperative research with universities.

Research cooperation has generally resulted in satisfaction to all participants, for several reasons. University and industrial research collaborators combine complementary skills and facilities. The university researcher generally performs more of the theoretical tasks, and the industrial researcher more of the sample preparation and characterization tasks. They perform the experimental tasks equally.

Industry-University Cooperative Research Centers
A second programme is the Industry-University Cooperative Research Centers programme, which stimulates industrial support of university research through the establishment of centres to create long-term collaboration between the university and industry in research areas of high mutual interest. This NSF programme, with co-funding from groups of industrial companies, initiates university research centres that are responsive to industry's research needs yet compatible with the research objectives and standards of the university. Industrially relevant research at the university results in

the exposure of faculty and graduate students to industrial interest and needs and provides a valuable funding source for the university. Concurrently, industry benefits by its exposure to university research capabilities and interests, by leveraging its funding *vis-à-vis* other participants, and through contact with staff and graduate students for possible recruitment purposes.

Industrial support and utilization of university research is a major programme goal. Research programmes of the centres correspond to the university's scientific and engineering areas of expertise. The centres research needs. All centres must increase the industrial support for their research programmes, as NSF support is phased out within a period of five years. A centre is considered a success when its research funding is at its original level or higher and NSF no longer provides support.

A key to the programme's success is the fact that the government (or, more specifically, NSF) does not determine the area of science or which industrial sector is to be addressed. It is responsive to the marketplace of collaborative research needs and ideas. The programme emphasizes local autonomy and separate development; each centre develops along its independent path determined by the university capabilities and interest and industry's needs and interest.

In order to avoid commitment of scarce programme funds to a centre concept that is not fully developed, NSF requires the university to start with a one-year planning grant, during which time they study and develop their research and management plans and, at the same time, determine the industry's research interest and willingness to support the proposed centre. Each centre that has been started or is in the planning stage has
the scientific disciplines involved. Thus the centre programme augments the scientific research support thrust of other NSF programmes.

After successful completion of the planning activity, a second proposal must be submitted by a potential centre for an operations grant. This second grant could cover an additional four-year period with NSF funding totalling up to $750,000, but gradually being reduced during the four years. At the conclusion of the NSF funding, the centre would be self-sustaining and entirely funded by industry. Currently NSF is funding or has funded twelve operational centres in such areas as polymers, computer graphics, welding, robotics, telecommunications, ceramics, materials handling, automation technology, hydrogen technology, and dielectrics. In addition there has been funding for eight planning grants in the areas of toxic waste, analytic chemistry, clean room technology, steel processing, tribology, electro-optics, monoclonal antibodies, and steelmaking. By the end of 1984 there will be about 20 operational centres, with an additional 8-10 active planning grants.

Critical elements

The most critical element in the university's management role is that of leadership. The centre leaders must be recognized scientists who have entrepreneurial talent and the drive to work with the university management and the marketing ability to work with industry representatives. Leadership ability is also required to attract university faculty and graduate students to take part in industrially relevant research.

The university interface with NSF and industry is one that changes as the role of NSF diminishes and the role of industry increases. As new companies are added and the industry interest solidifies, the universities must be responsive to industry by aligning their research goals within limits they feel are appropriate.

Industry's financial support is a major goal of the NSF programme. We see an increased awareness on the part of individual companies of the value of pooling their research dollars with other companies. In a typical centre they have access to $600,000 worth of research for a $30,000 investment.

Center membership requires more than financial support, however. It requires active participation in the centre by industrial leaders. They must provide qualified researchers to monitor the scientific progress of the research and top-level managers to act as advisors on management and operation of centres. It is this industry management role which guides the research and determines the industrial relevance.

The process of initiating and operating a centre is a complex social process, and the success of a centre is more than just a function of scientific merit and dollars. All of the thriving centres are both centres of scientific excellence and healthy social organizations with regard to industry-university activities. A great deal has been learned about centres and industrial support of them. A practice manual based on this experience has been published together with a set of case studies. These documents have been widely distributed and are used as models by universities and industry.

Accomplishments

The accomplishments of the centres programme have been many. They have led to stronger and more effective interdisciplinary university research programmes with industrial relevance. The amount of NSF and individual company funds required has been minimal. The students coming out of the programme are well rounded and ready to assume their roles in industry upon graduation. The most important and long-term result of all these accomplishments is the coupling of fundamental research to technological utilization.

Benefits and problems of cooperative research programmes
There are some obvious benefits of cooperation between university and industry researchers. First, these programmes leverage federal funds. Overall industry funding of university research now stands at less than 4% of the total funds, compared to nearly 70% for the federal government. Anything that can be done to increase industry participation is in the national interest. Second, industry-university cooperative research programmes provide continuity of funding, strengthen academic departments, and enhance graduate student and post-doctoral training. Third, industry-university cooperation improves job opportunities for graduates of the participating universities, while at the same time enhancing the transfer to industry of the knowledge embodied in these graduates.

From our experience, several problems face universities, industry and government in the development of university-industry cooperative research. One is the possible restriction of information exchange on progress in research when dealing with proprietary issues of industry. Another concerns the ownership of patents and the nature of licensing agreements. The foundation is experimenting with and evaluating various modes of industry-university interactions to better understand these concerns.

In the short term, our experience has been that issues such as academic freedom, publications and patents can be dealt with through open discussion between industry and the university and do not require government action or intervention. There is a great deal of discussion and concern about these issues as relationships are being established, but they tend to diminish once the research begins. While the issues are important, the industry-university partners seem to be able to accommodate the interests of each in a way which contributes to the public good.

Small business innovation research program

Background
Small high-technology firms represent an important national resource. Not only do they have strong research capabilities but

they represent one of the best mechanisms to convert research into technology. Such firms accelerate technological innovation and can increase the commercial applications of research for its potential economic and social benefits.

At NSF, the Small Business Innovation Research (SBIR) programme started in the fiscal year 1977 when Congress directed NSF to award $1 million of research funds to small business firms. The programme has grown to a level of $5.5 million in FY-1983 and will more than double again by FY-1985 to a level of over $12 million, which will be about 1% of the NSF research budget. This large increase is due to the legislative requirements of the Small Business Innovation Development Act of 1982, which prescribes SBIR funding levels which will reach a government-wide total of about $500 million in ten agencies by FY-1987.

Objectives
The objectives of the Small Business Innovation Development Act of 1982 are to:

- Stimulate technological innovation;
- Use small business to meet federal research and development needs;
- Foster and encourage participation by minority and disadvantaged persons in technological innovation; and
- Increase private sector commercialization of innovations derived from federal research and development.

SBIR provides an important opportunity for small science- and technology-based firms to participate in federally funded research. It enables small firms to propose longer-term, higher-risk creative ideas in government programme areas which could only be pursued by them with outside support. These awards have also been valuable to many firms in their obtaining private investment, other R&D awards and working arrangements with universities, government agencies and large companies.

Structure
The federal programme is structured in three phases and is modelled after the NSF SBIR programme. In general, Phase I supports research for six months for up to $35,000 on important scientific or engineering problems; Phase II funds those projects found most promising after Phase I for up to $200,000 for one to two years; and Phase III is the product development phase to pursue

the commercial applications from the government R&D conducted in Phases I and II. It is entirely funded with private capital, usually from a venture capital firm interested in investment or a large industrial company looking to license or purchase new technology.

The programme design provides a stopping point to evaluate projects after Phase I to determine whether a large Phase II investment should be made. Stopping for evaluation after Phase I results are known has received high marks from some who evaluate R&D policies and practices.

The programme is extremely competitive, with almost four thousand proposals received and less than five hundred awards made to date at NSF alone. Proposals have been received from all 50 states and awards have gone to firms in 36 states and the District of Columbia. Historically, only one out of eight proposals have been funded in Phase I and only one out of two or three of these receive a Phase II award.

Small high-technology firms often have excellent research capabilities in industrially relevant areas and frequently in places where possible small markets may deter larger firms from investing. Advanced instrumentation research for scientific and industrial measurement is an example of this. Phase II awards have included a wide range of interests which may lead to new scientific or industrial instruments such as tomography, spectral correlators, ion microscopy, and laser-based photo-acoustic measuring.

About one-half of NSF SBIR awardees have affiliated with universities, mainly using university scientists and engineers as consultants, but occasionally involving subcontracts or the use of specialized university facilities. About 10% of the budgets of successful NSF Phase I projects have been earmarked for such activities. Although this may not be representative over all of the agencies, a significant amount of the total SBIR funds may be spent at the nation's colleges and universities.

NSF suggests that small businesses develop ties with universities in responding to the annual programme solicitation. University researchers are often more familiar with the "state of the art", as well as having more experience in proposal writing. If a small business uses this capability, it should result in a better written, more credible, and technically accurate proposal, and can significantly increase its potential for an award.

Most of the university-industry interaction occurring in SBIR projects is in a consulting arrangement. Some projects have utilized the unique equipment available in universities, and a few have subcontracted research to the university. However, many small businesses have little or no contact with the university. Because of the advantages of universities in responding to the SBIR programme, this is changing.

Benefits

Since a small business has everything to gain and the effort required by a university researcher is relatively small, there seems to be good reason for cooperation between these two groups in preparing a proposal. With the congressionally mandated growth of small business R&D funding, the opportunities and importance of interaction with small business will increase. More projects with higher funding will be available from multiple federal sources. In these projects, this type of university assistance to small businesses can form a basis for regional development. In fact, states are already in the process of developing programmes in which the university will assist small business in their region to identify and respond to the programmes initiated by the Small Business Innovation Development Act of 1982.

SBIR offers a unique mechanism to leverage federal research in many different areas. In addition to those previously stated, SBIR firms increasingly invite the authors of publications, who are often doing basic research, to be consultants in their projects. This approach couples basic to applied research and may provide a practical way for more ideas to move from the university into applications and the economy, to introduce university scientists to high-tech firms and to encourage joint research efforts.

Large firms leverage the federally funded SBIR research. They are very interested in the advanced high-risk research funded by SBIR. Follow-on funding commitments have been received from many firms such as Sohio, Arco, Ford Aerospace, Fairchild, and many others to support Phase III efforts. Some large firms see the SBIR programme as a farm system to stimulate innovation in large firms through their licensing or acquisition of the technology generated by small firms. Venture capital firms also leverage SBIR research many times.

Finally, SBIR leverages small business start-ups which are important to economic development. Of these grantees, 45% had ten or fewer employees when they submitted their Phase I proposal, and 13% of all awards have gone to minority-owned firms. This certainly represents a major step in utilizing a wide range of important national resources.

Financial

In the fiscal year 1984, a total of $16.6 million was allocated for these programmes: $7 million for cooperative research projects; $3 million for cooperative research centres; and $6.6 million for small business innovation research. In addition, $1 million will be spent for productivity improvement research.

The funding history for the various programme elements shows a relatively constant level of support. However, this can be misleading in the true impact of funds leveraged from other sources. The cooperative project programme's real effort is almost three times the amount in the NSF budget. The cooperative centres programme leverages industrial resources at a level of four to five times the amount provided through the NSF budget. Follow-on funding due directly or in part to SBIR support now exceeds $50 million, primarily related to the first solicitation which has completed both Phases I and II. Leverage in this programme is about five times the NSF support provided. In total, the NSF annual investment of about $16 million in industrial innovation related programmes results in direct industrial support of over $60 million dollars or an overall ratio of about four to one.

Future plans

Reporting to the President of the United States, the Business-Higher Education Forum observed that "industrial competitiveness increasingly depends on the speed with which companies can apply the latest scientific advances. This fact alone argues strongly for increased university-industry cooperation at the earliest stages of research" (see *America's Competitive Challenge: The Need for a National Response*, report of the Business-Higher Education Forum, April 1983).

The report suggests that in addition to existing programmes, such as those at NSF, it could be beneficial to expand government efforts to encourage participation in cooperative research by business and educational institutions that have not been substantially involved previously.

NSF is one of the important actors in the science and technology policy arena affecting academic science and engineering, and definitely is looked upon as exercising a leadership role in the national trend of industry-university cooperation in research. The IUCR projects and centres programmes have been the primary methods utilized so far and NSF has provided the leadership. This leadership is being continued by providing a focal point for reliable information and guidance to academic institutions, industrial firms and other organizations in developing cooperative research linkages.

We are also proposing initiatives to maintain the NSF leadership in promoting industry-university cooperation with efforts to bring into the research community an underutilized resource — non-traditional institutions, such as small colleges and universities, small business, and trade associations. This effort is consistent with recommendations of the Business-Higher Education Forum report to the President. Our staff makes numerous presentations at university and small business sponsored seminars at which information describing programme opportunities is presented. In addition, we will continue to meet with small college faculty groups to provide information on cooperative research opportunities in the projects and centres programmes.

JAPANESE INDUSTRIAL POLICY:
COMPETITION AND COOPERATION

T. Kobayashi

When I was in Japan I belonged to MITI (Ministry of International Trade and Industry), and some of my work had particular reference to the machinery and electronics industries including the computer industry. So on this occasion I would like to explain the attitude of people within the Japanese government (particularly MITI's attitude) to policy making for industry, including a case study of computer projects.

First, this paper gives a rough outline of the concepts behind Japanese industrial policy, and second, by means of a case study, it shows how this policy works in the computer industry, with an explanation of some details of the fifth generation computer project which is now going on in Japan.

Japanese industrial policy in general

Aims

First I want to explain MITI's general industrial policy. Any such industrial policy in Japan is faced with two major tasks. One is to facilitate the development of high technologies, and the other is to encourage adjustment to industries facing structural problems.

1. High tech: Let me explain about high technologies in Japan. The history of Japanese growth after the second world war is mainly concerned with catching up with the advanced countries in Europe and the United States by studying their industrial results and implementing them. But now the R&D situation worldwide has reached a difficult stage for discovering the seeds of new breakthrough technologies. By this I mean that now major industrial activity using conventional R&D is producing only minor changes on previously obtained results, so that a radical alteration in type of R&D activity is needed in order to produce the necessary revolutionary technologies. I think that it is essential for the future development of industry that we achieve the new breakthrough

technologies which will bring industry to a highly sophisticated level. As one of the major advanced countries, Japan will have to make her contribution to the discovery of these revolutionary high technologies. However, there are some major difficulties for carrying out such R&D. For example, this R&D requires a very long-term view, a great deal of money, time and effort, and in addition the anticipated results are often vague and the value of such results are difficult to estimate. So, government must act as a leader in promoting this type of R&D, and there must be international cooperation between governments and between industries as well.

2. Adjustment: Another big task for MITI is to encourage adjustment in industries facing structural problems. This task is completely different from the one I mentioned before, but it also has a great part to play in making the industrial structure more sophisticated. This means that there are some industries which are facing difficulties because of radical changes in economics caused by the oil crisis or other similar influential events. Examples of such industries are the petrochemical and the aluminium industries in Japan today.

Tools
MITI has four main tools with which to realize these major tasks.

1. Visions: One such tool is to create visions, either for whole industries or for each individual industry. Usually each vision has three main parts: firstly to analyse the changing circumstances around the industry; secondly, on the basis of such analysis, to clarify the trends of economics and the industrial structure; and finally to indicate how the industry can change its industrial structure and proceed in the direction it desires. Creating such visions is not only the role of government. The industrial and academic sectors are also involved in such discussions, and exchange opinions with each other.

2. Financing: The second tool for MITI is government financing. Of course the main investment for industry is decided by the private sector and the market mechanism will provide money for investment. But there are investments that need large amounts of money and include high risks and long lead times. For such financing, there is the Japan Development Bank.

3. Tax incentives: The third tool is tax incentives. There are tax incentives to achieve certain policy goals such as securing stable, long-term energy supplies, promoting economic cooperation with the developing countries and encouraging technological development.

4. Budget: The fourth tool is budgetary appropriations. The general budgetary appropriations are fairly small and are designed to draw industry's attention to the proposed direction it should take by encouraging technological developments in specific areas.

A typical example is the "Basic Technology for Future Industries" scheme which is a major project for promoting high technology and was begun in 1981 as a result of a "vision" that was compiled in 1980 on the future of industrial policy in Japan. In that vision it was recognized that Japan, relying so heavily on external sources for her raw materials, must make her contribution to the world economy by developing high technologies. Therefore, we chose several technological targets requiring a revolutionary and original approach. But such technology also requires a long-term plan involving vast quantities of money and carrying a high risk. Therefore it is not possible for private industry to carry such a burden alone. In addition, such a scheme needs cooperation between the government, private and academic sectors.

Under these conditions we chose twelve themes in three major fields. These fields are new materials, biotechnology, and new electronic components. Almost all these projects are planned to take ten years. Each project is divided into three stages, each of which has its own target. At the end of each stage the results are evaluated, external circumstances worldwide are reassessed, and the remaining targets modified accordingly. In this way the flexibility of the scheme is maintained.

Summary

To conclude the first part of this paper, I will summarize the aims of Japanese industrial policy. These are: first, to find possible and desirable directions for industry to take; second, by discussions with all those involved, to create a cooperative atmosphere in which to promote a unified vision and to decide on the consequent R&D targets; and third, on approaching these targets, to maintain competitiveness and vitality within each industry.

Case study of the computer industry

As the second part of this paper I will explain, as an example of Japanese industrial policy, the government's measures concerning the computer industry.

History

Before going into detail about the fifth generation computer project itself, I want to explain the Japanese government's scheme regarding the mainframe computer companies. The history of this involvement is recent, covering only the last ten years.

1. Computer scheme: The first scheme, called the "Computer Development Promotion Scheme", began only in 1972. It was begun because in the summer of 1971 the Japanese government decided that eventually the Japanese market for computers and integrated circuits would be made free for investment and imports. The government and MITI thought that before the opening up of the market the Japanese computer industry must be given time and encouragement to catch up with the rest of the world, in terms of competitive technologies and development abilities.

Therefore, a new five-year scheme for the mainframe computer industry was begun. At that time (1972) there were six mainframe computer companies in Japan. For the purposes of this project, MITI divided them into three groups. One group consisted of Fujitsu and Hitachi, that is, IBM-compatible computer manufacturers. Next came NEC and Toshiba, that is, non-IBM-compatible computers. The third group was Mitsubishi and Oki, that is, smaller mainframe computer companies. This scheme was brought to a successful conclusion and these three groups came to produce an extremely competitive computer series. During this time, at the end of 1975, the Japanese market in computers was freed, both for investment and for imports.

2. VLSI scheme: The next scheme, called the "Very Large Scale Integrated Circuit Development Promotion Scheme for the next generation computer" (or, more simply, the VLSI scheme), started in 1976. This was a follow-up project to the initial scheme. At this point the main items of hardware technology to be considered for the next generation computer were the VLSIs, and in this project the Japanese companies were to develop a fine processing technology for VLSIs, of the submicron order. At this stage one of the Japanese mainframe companies (Oki) dropped out of the project, so MITI made Mitsubishi join the Fujitsu and Hitachi group, thus making two groups, one IBM-compatible and one non-IBM-compatible. As this VLSI project concerned the vital component technology it had to be completed quickly, and in fact only took four years, from 1976 to 1979.

3. Basic technology scheme: The next scheme, called the "Basic Technology Development Promotion Scheme for the next generation computer", was divided into two parts: first (and this is the main part of the project), the operating system or basic software technology, and second the peripheral and terminal technology. These items were chosen because the new generation computer

would be characterized by decentralized processing, including a network system and database, and also by the application of the Japanese language, so that the basic software (a weak point in the Japanese computer industry at that time), and the peripheral and terminal technology were important. The R&D into the basic software was conducted by the two groups mentioned above, and that into the peripherals and terminals by eight private companies — the five from the above groupings plus Oki, Matushita and Sharp. This five-year scheme started in 1979 and was due for completion at the end of the 1983 financial year.

The results of the first computer development generation scheme were the 3.5 generation, or LSI computer, and the establishment of the whole Japanese main computer series. The next scheme (VLSI) concerned component technology and continued with the operating system, that is, the software. These technological developments are being implemented as a new computer series, which will be practically a fourth generation computer series.

4. Fifth generation computer scheme: The fourth scheme is the "Fifth Generation Computer Project". Preliminary investigations for this scheme were carried out in 1981 and the ten-year project itself started in the summer of 1982.

MITI thought that although at that time the Japanese computer industries were competitive and held a good position worldwide, a new breakthrough in technology must come soon and that Japanese industry must play an important part in any such breakthrough. Therefore, MITI set out to find what kind of technology was needed for the future computer industries. The result was the concept of a fifth generation computer which has a completely different structure or processing method, moving towards finding artificial intelligence.

5. Analysis of the history of the schemes: Thus, the salient points of the Japanese government schemes over the past ten years have been:

First, the grouping, which was designed to maintain competitiveness. Of course, we must compete with foreign companies such as IBM but other companies are also necessary, even in Japan — if one company has a monopoly, it may fall into the trap of becoming lazy and depending more and more on the government for support. If it has rivals it will strive to be successful and competitive.

Second, the joint research centre, which started from the VLSI project. At first it was difficult for personnel from different companies to work together in one building, but after about a year the atmosphere in the centre improved rapidly and a frank exchange of information between companies led to good results and complete success for the VLSI project.

Third, there is the matter of the direction of government support. The first scheme from 1972, the computer development promotion scheme, was partly a result of foreign pressure on Japan to free her market. So the government subsidy was a general one designed to accelerate the whole of the R&D process. But the next two schemes (from 1976 to 1983, thought of as one eight-year project) were to boost the weak points of the Japanese industry, i.e. VLSIs and basic software, and the subsidy was given only to these areas. By contrast, the new fifth generation computer project is designed to enable Japanese industry to play its part in breaking technological barriers.

Details of the fifth generation computer scheme

Background
One of the reasons for this scheme is the movement towards a knowledge-based society. Economic development has led to "knowledge" becoming the new resource. Consequently, the production and processing of this new resource is an important part of future economic development and the computer is one element that will assist such production. In this sense, application areas for the computer should widen rapidly.

However, we have at the moment three problems with computers. One is that they are designed for numerical calculation and are not well equipped to deal with non-numerical information such as language or pictures. Secondly, the computing method is only a one-by-one system (however fast it may be), and the memory is only one-dimensional. That is because when computers were first developed, hardware costs were very high, which led to the very simple nature of the hardware structure. But this simplification has, in turn, produced the third problem, which is the increasing cost of developing the software for what are basically software-dependent systems. This is called the "software crisis".

Analysis of the possibilities of the project
The fifth generation is called "fifth", but is in fact based on a completely different concept from all those computers at present in existence as it uses non-numerical data and a new structure. So basically it is a knowledge information processing computer with such core functions as problem solving and inference systems and knowledge base systems which cannot be handled within the framework of conventional computer systems.

At present we possess both the potential and also the requirement for the development of the fifth generation computer.

(a) VLSIs have led to reduced hardware costs, so that an almost unlimited supply of hardware is available.

(b) It is likely that we are nearing the limit of potential for development of the rapid electronic element. In addition, new architecture for parallel processing seems to be one solution to the problem of how to achieve speedier computing.

(c) Parallel processing has the advantage of using many VLSIs which are already produced at a cost effective level, thus making parallel processing a practical proposition.

(d) The basic functions necessary to realize artificial intelligence, such as the processing of non-numerical data, and inference and learning, are at a very low technological level in our present computers.

Time schedule and budget

The Japanese fifth generation computer project, which started in 1982, is due to run for ten years. This period is divided into three. The first part, taking three years, is to be used to establish basic technology for knowledge information processing. The second part will take four years and will be used to check and evaluate, by a systematic approach, the basic technology that was produced in the first period, and to gain an overall view of the knowledge information processing system. The third and final period is for three years from 1989. It will produce the prototype of, and act as a feasibility check for, the actual implementation of the total information processing system.

However, this will be a primitive, non-commercial prototype. It is hoped that after that, private companies will take up the responsibility of commercialization.

Before the commencement of this project the expected total cost was estimated as 100 billion yen (250 million pounds) with 10 billion yen (25 million pounds) for the first part, 50 billion yen (125 million pounds) for the second, and 40 billion yen (100 million pounds) for the third part. The cost of the first stage was 100% funded by the government, and will be about 15% less than the sum estimated. This is different from previous schemes where the government subsidy was less than 50%. However, the percentage of government support for the subsequent stages has not yet been decided. It will depend on the Japanese budget system which allocates funds on an annual basis.

Structure of the project
The Institute for New Generation Computer Technology (ICOT) was established to promote this project. ICOT has 50 employees, including 40 engineers seconded from the eight companies mentioned above and from the NIT (Nippon Telegraph and Telephone Public Corporation) and the ETL (Electrotechnical Laboratory, a government laboratory).

Present situation
The first period of the scheme, which deals with basic technology, is divided into four R&D subjects. These are the inference subsystem, the knowledge base subsystem, the basic software system, and the pilot model system for software development.

We have now finished the second year of the first stage of the project. Last year the concept of each subsystem of the project was realized in principle, and the first pilot model computer will be functioning shortly.

One other activity connected with this project is international cooperation. We feel that as this project is aiming at technological breakthroughs, this is an important part of our work. ICOT is trying to communicate with other similar projects for mutual benefit, and at an initial stage invited several top rank scientists to discuss with them a wide range of topics within this field.

Conclusion
The case study I chose as an example of the high technology industries was the computer industry, but it is in fact very difficult to predict good results within any high technology industry simply because extreme patience and a sustained effort over a long period is required. Therefore it is my belief that international cooperation is necessary if anything substantial is to be achieved. We do have some good experience of international cooperation, for example in the field of energy production, but where computer development is concerned, although the schemes are government supported, the main impetus still comes from the industrial sector, and to date there are no good examples of international cooperative ventures between two major high technology industries (as opposed to links between individual companies). Therefore, I believe we must find some good schemes for cooperation between such industries where our future efforts towards high technology R&D may be combined. The most important point of these schemes is how to harmonize cooperation and competition in one project. It is with this in mind that I have a keen interest in the success of the European Esprit project.

A RESEARCH LABORATORY AS AN INNOVATION POLICY INSTRUMENT

Erkki Ormala

Different innovation policy measures and their instrumentality have been widely discussed during the last few years. Rothwell and Zegveld have presented a list of alternative government innovation policy tools and analysis of policy recommendations by type of tool in some countries (Rothwell and Zegveld, 1981). There is no general agreement about the applicability of these tools. Goldberg has concluded that there is little or no scientific evidence to support the existence of any goal achievement relationships and efficiency of innovation policies and measures (Goldberg, 1981). This negative outcome cannot be interpreted as a statement to the effect that innovation policy tools are useless and thus should be discarded. On the contrary, there are clear indications of increasing government intervention in the innovation system in most industrialized countries (see, e.g., Blume, 1981; OECD, 1983).

In most industrialized countries the scientific and technological infrastructure plays an important role in the performance of national research and development work. The strengthening of this infrastructure and its more effective use in the enhancement of innovation activity seem to become even more important in the future. Non-profit research laboratories are one of the main policy instruments and they can be found in almost all industrialized countries. These research laboratories are organized and financed in a variety of ways in different countries (see Rothwell and Zegveld, 1981). They also have varying responsibilities which may include, among others,

- collective industrial research
- technology transfer
- contract research for a single firm
- contract research for the administration

Though there are examples of successful contributions to innovation activity by the research laboratories, there are also examples of failures. The factors associated with success and the mechanisms of the interaction between research laboratories and the innovating industry are not known.

In Finland the main features in the development of the production structure have during the last twenty years been the diversification of the metal industry and the strong development of the electronics industry. However, the production structure of Finland is still rather one-sided and thus easily vulnerable. Also, a rather small part of industrial production is research intensive compared with other industrialized countries, and particularly the West-oriented exports of Finland still emphasize capital-intensive and relatively few processed products. These facts have been important starting points in the discussions in Finland on the acceleration of technological development and the importance of governmental technology-political measures.

According to several policy statements in the early 1980s, research funding will be increased rapidly and great efforts will be made in order to attain a more rapid growth than the average increase in expenditure, due to the nature of research and development as a creator of innovations. This growth will be based on an increase in funding both in the private and public sectors.

The share of technical research and development of gross domestic expenditure on R&D (GERD) was in Finland in 1981 nearly 60%. This was divided between the three main parts of technical R&D in the following way: business enterprise sector 80%, The Technical Research Centre of Finland (VTT) 15%, and technical universities 5%.

This paper tries to illustrate what the role of a public electronics research laboratory can be, and what are the mechanisms in the research laboratory/industry interaction. The paper relies on the results of research carried out at VTT. This research was aimed at studying the research performance and quality of four of VTT's laboratories. This subject has been widely discussed during the last few years at VTT and some of the conclusions have been reported earlier (Lemola et al., 1983; Ormala, 1983).

The decision to use VTT's electronics laboratory as an example is based on two factors. Firstly, the electronics laboratory represents high technology, which has become one of the main fields of technological development in many countries. Secondly, the electronics laboratory is located in a developing area in Finland, and it is important to know whether a research laboratory can succeed outside the main industrial areas and how the laboratory can contribute to the areal development.

The Technical Research Centre of Finland (VTT)
VTT is one of the main innovation and research policy instruments in Finland; it is a government research organization covering a diversified and wide range of research work. Its purpose is to create, maintain and develop the technical knowledge needed by the Finnish economy and administration. Its main activities are research and development work, testing and inspection work and information services. Priority is given to applied research related to concrete problems and needs.

VTT's 31 laboratories cover most fields of technology. With its broad range of operations, VTT is in a good position to combine the expertise of several fields in resolving a single problem. The most important fields of research are building and community technology, materials and processing technology, energy and information technology. The number of personnel is about 2400 at the moment.

VTT carries out research work on its own initiative with financing from the state budget and contract research for administrative bodies and industrial companies. Contract research already comprises about 65% of VTT's activity; it is mainly charged for at cost.

In Finland VTT has quite an important role both as a typical research institute and as a service organization. VTT influences innovation both indirectly by strengthening the technological infrastructure and directly by supporting the innovation projects of individual companies. Its role in so far as its direct impact is concerned can be summarized as follows: VTT develops products and processes, usually having been commissioned to do so by companies. From the point of view of the companies, these are generally (whether large or small) parts of a wider innovation process. So far, it has been rare for VTT to take responsibility for the whole innovation process. The development or innovation work done on VTT's own initiative has been primarily development of instruments for VTT's own use. It is improving the technological infrastructure by performing its other tasks.

The main advantages of VTT in innovation activities are firstly its wide scope, and secondly its good relations with industry and with technical universities. The structure and organization of VTT enable the rise of a very broad base of knowledge within a project, using expertise from different fields. VTT is somewhere between the universities and industry; this enables contact with both sides (technological possibilities and market needs) which is a necessity in innovation (see, for example, Sahal, 1981; Freeman, 1982). VTT gets information from industry as to its needs, while good relations with the world of basic research make available knowledge of new methods. VTT also has wide international contacts, both with industry and with research institutions.

Electronics laboratory
The electronics laboratory (ELE) of VTT was founded in 1974 in Oulu, about 600 km north of Helsinki, the capital of Finland. Orlu is the administrative centre for Northern Finland and it is also an industrial city. In 1982 the research staff of the laboratory numbered 92. At the beginning of 1983 12 employees joined a new electronics company as a result of a technology project of the laboratory financed by 14 electronics companies.

The total budget in 1982 was 12.2 million FIM (about US$2.2m). Industrial customers were responsible for financing about 70% of the laboratory's activities. The remaining finance, which is concentrated on applied research, comes from the state budget and administrative bodies. The laboratory concentrates on research and advanced project development in measurement technology, machine automation, electronic manufacturing technology and computer aided engineering.

Northern Finland is almost entirely a so-called developing area, and one of the reasons for locating ELE in Oulu was to support the industrial development of the area. The rapid growth of the laboratory and its ability to raise external financing (higher than the average at VTT) have shown that locating ELE outside the main industrial area in Finland has not been an obstacle to its success.

Another way of looking at the areal impact is to compare the areal distrubtion of ELE's external financing from the industrial companies with the areal distribution of the manufacture of fabricated metal products, machinery and equipment, which are the main industries served by ELE. In Figure 1 these distributions are given. The country is divided into three parts: Oulu district, Helsinki district, and the rest of the country.

Both Helsinki district and Oulu district seem to be over-represented in the financing distribution. This can be partly explained by the concentration of the electronics industry in the Helsinki district, but that does not explain the share of the Oulu district. Consequently the location of the laboratory really seems to have some impact, though not totally clear, on local industry. Another explanation of the difference may be that there are very good communication channels between the capital and Oulu, but

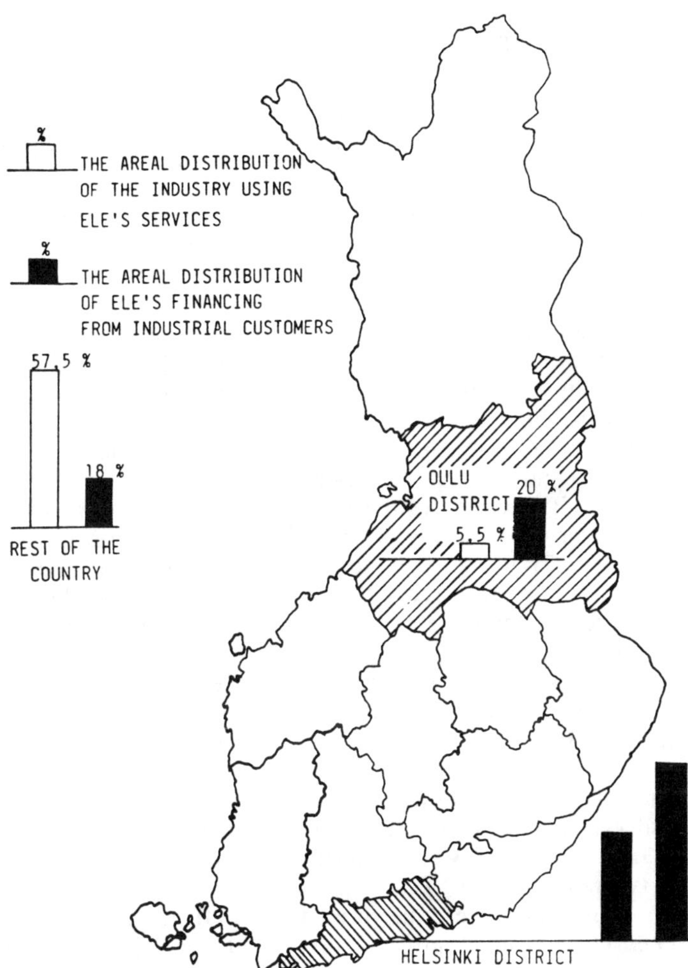

Figure 1. The areal distributions of the industry using ELE's
services and ELE's financing from industrial customers

that is not necessarily the case between Oulu and the rest of the country.

The number of small and medium sized companies using ELE's services seems to be bigger than the number of large companies (see Table 1). But the financing from the large companies is substantially bigger than from the small and medium sized companies. The size of the customers does not explain the differences in areal distribution.

ELE has traditionally published more in international and Finnish professional journals than in VTT's own research report series. The annual averages have been 7 international contributions and 13 national papers. On average, ELE publishes less than VTT's laboratories, and this is at least partly explained by the strong orientation of the laboratory towards product development and confidential research contracts.

Patent statistics show that two patents have been granted and three applications are still being considered by the authorities. Some patents have also been granted to individuals on the staff. Generally, patent statistics are not a very good measure of research performance because the patent rights are transferred to the customer in contract research projects.

VTT mainly recruits employees directly from universities. A typical career of a research engineer is such that after a few years (perhaps 3 to 7) he goes into industry. This is one of the important contributions of a research laboratory. After having worked with new technology for some time, the research engineers are highly qualified experts in their own fields and can bring the newest technological know-how and practical experience of its applicability with them into industry. Every year on average five experienced research engineers have left ELE for industry.

Evaluation of research projects
In order to assess VTT's research performance and quality, an evaluation of four of VTT's laboratories was carried out. One of these laboratories was ELE.

The evaluations were performed by evaluating a restricted number of completed projects (about 30 per laboratory). The projects were evaluated by persons who were supposed to have the best ability to evaluate the relevance and quality of the projects. Publicly funded "general" projects were evaluated by a steering committee member who represented the users of the results of the projects. Contract projects were evaluated by a representative of the contractor. All respondents were personally interviewed. A

Table 1.   The number of industrial customers and the
financing from the customers in different
firm size categories

Number of employees of the customer companies	1-49	50-499	500-
number of customers	9	14	17
financing (1000 FIM)*	476	493	4642
financing/customer (1000 FIM)*	52	35	273

*US$1 = 5.4 FIM

questionnaire was designed to give evaluations on the objectives which reflect the relevance, quality and economic result of the projects. Test interviews revealed that different sets of questions had to be used to evaluate the relevance of contract projects and general projects.

It is important to define the two project categories: (a) general projects which are public and at least partly financed by public funds; (b) contract projects which are confidential and completely paid for by the customer.

The samples of about 30 projects per laboratory were random. Consequently the numbers of general projects and contract projects in the sample represented the mutual proportions of these projects in the laboratory. In ELE's sample there were 8 general projects and 17 contract projects. The projects had been completed either in 1980 or 1981. The total number of completed projects was 67.

The average size of the projects turned out to be smaller than was expected. Both the general projects and the contract projects were about 1.5 man years (260,000 FIM) on the average. Yet ELE's contract projects were considerably bigger than the contract projects of the other laboratories. The general projects were of the same size in all laboratories.

All ELE's general projects were concerned with developing or transferring new technology and all contract projects were product or process development tasks which were parts of development projects of the customer company. Both the general and the contract projects were much more diversified in nature in the other laboratories.

The costs of ELE's contract projects were on average about 42% of the total costs of the customer project. This share in ELE was higher than in the other laboratories. The relatively high figure suggests that ELE can take more responsibility for development than had been considered possible. When looking at the distribution of this share, one could find examples of projects where practically the whole development of a new product was carried out by the laboratory.

The respondents were asked the reason why the project was given to VTT. As to the general projects the main reasons seemed to be that the idea was generated by ELE (6 answers) or that the results were interesting to many companies and VTT as an impartial organization was considered proper (3 answers). The contract projects were given to ELE because ELE had superior know-how (9 answers) and required facilities available (7 answers), or because the customer lacked its own resources (9 answers). These two reasons seemed to be dominating and almost equally important. The results were almost the same in all laboratories. The services of a research laboratory are used if the laboratory can provide superior know-how or available facilities that are worth buying. The companies also seem to use the services to balance an overload of their own research resources. Both small and large companies seemed to justify their use of the services in the same way.

Only two of ELE's general projects and three of the contract projects were such that the respondent could not see any alternative means of doing the research. The most often mentioned alternatives were the technical universities (7 answers), consulting companies (7 answers) and the customer itself (12 answers). However, the respondents seemed to be uncertain about this question. Many "perhaps" answers were given. Also the reliability of this question turned out to be relatively low.

The impact of ELE's general projects

No objective way to measure the impact of the general projects was found. Consequently the relevance of these projects was assessed only on subjective scales and the analysis remained comparative and rather descriptive. The respondents were asked to assess the national economic utility and the scientific/technological utility of the projects. The assessment was performed by using a set of questions (see Table 2). The laboratory directors were asked to assess the utility of the projects to their own laboratories.

The assessment generated a kind of profile for each laboratory and a subjective estimate of the total national economic utility. The form of the profile was much more interesting than the numeric values. Each laboratory had a unique form of the impact profile and the comparisons between them give a visual description of the orientation of the laboratory. Similar profiles were assessed for the scientific/technological utility and for the utility to the laboratory.

Four of ELE's general projects had succeeded in achieving their goals. But it was too early to evaluate the success of four projects. The respondents were also asked to name the most important users of the results of the projects. The companies represented in the steering committee, other companies and VTT itself were considered the most important user groups. In this respect the results in all four laboratories were roughly the same.

The relevance of ELE's contract projects

The relevance of the contract projects was assessed by assessing the nature and importance of the customer projects, the importance of

Table 2.   The national economic utility of ELE's general projects

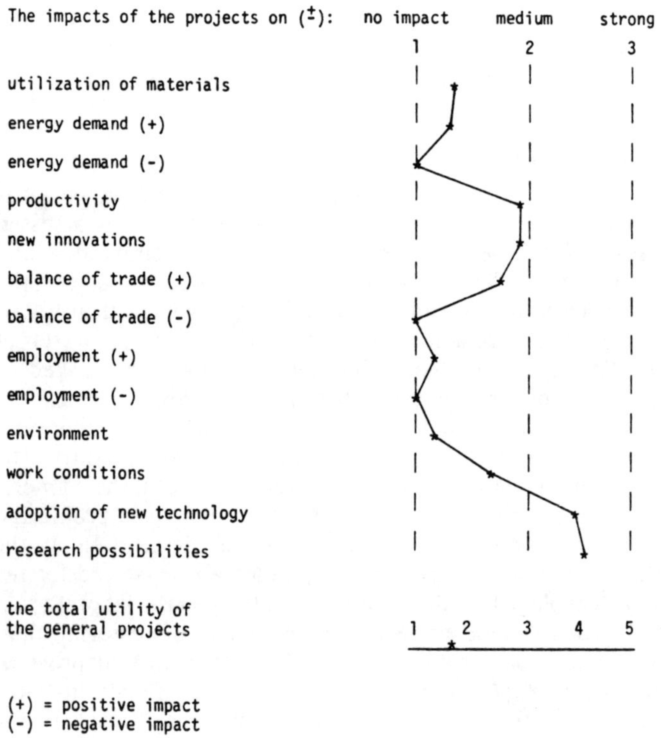

The impacts of the projects on ($\pm$):   no impact   medium   strong

	1	2	3
utilization of materials			
energy demand (+)			
energy demand (-)			
productivity			
new innovations			
balance of trade (+)			
balance of trade (-)			
employment (+)			
employment (-)			
environment			
work conditions			
adoption of new technology			
research possibilities			

the total utility of
the general projects   1   2   3   4   5

(+) = positive impact
(-) = negative impact

the contract projects to the customer projects, and the utility of the contract projects to the performing laboratory.

The utility of the contract project to the performing laboratory was assessed by using the method described in the previous section. The economic success of the customer projects, the importance of the contract projects to the customer projects and the expected sales of the customer projects are presented in Table 3. None of the customer projects had failed. Only one of the projects had succeeded only partly, and nine had turned out to be economic successes. The interviews were done in the spring of 1982, almost immediately after the contract projects had been completed. This accounts for the high number of answers indicating that success or failure could not yet be identified. In thirteen cases the contract project had been important or very important to the customer project. In four cases the importance had been moderate. There were no cases in which the importance of the contract project was considered small or irrelevant. This finding supports the result mentioned previously where the contract projects had a higher share of the costs of the customer projects than had been expected.

In some cases the respondents mentioned the expected sales of the products. The uncertainty in these figures is high and they should be interpreted only as an indication of the rough size of the product sales than as any kind of numerical measure of the performance of the laboratory.

The average expected sales were 2.6 million FIM per project. The average size of the contract projects was 260,000 FIM, which is 10% of the average expected sales. The average development costs of the products in total were about 600,000 FIM, which is roughly 23% of the expected sales, which is a fairly typical figure in the electronics industry.

When looking at the relevance of the contract projects of the other laboratories, some similarities could be identified. None of the customer projects had failed, and the proportion of uncertain answers was smaller than at ELE. This result suggests that the companies tend to perform their really risky projects themselves and use the outside research services only when the risk of failure is relatively low.

The importance of the contract project to the customer was lower in the other laboratories, which was also indicated by their smaller share of costs.

Conclusions

Research laboratories are used as innovation policy tools in all

Table 3. The relevance of ELE's contract projects

Proj. No.	Economic success	Importance of the contract project	Expected sales (1000FIM)
1	uncertain	very important	-*
2	uncertain	important	-
3	yes	moderate	25,000
4	uncertain	important	1,800
5	yes	very important	-
6	yes	important	-
7	yes	moderate	-
8	uncertain	moderate	4,000
9	yes	moderate	1,000
10	yes	important	-
11	uncertain	important	-
12	uncertain	important	4,000
13	partly	important	-
14	yes	important	5,000
15	uncertain	important	-
16	yes	very important	-
17	yes	important	10,000

* - indicates that the respondent did not mention a figure

industrial countries. However, there is very little evidence of their real contribution to innovation activity. The Technical Research Centre of Finland is a major innovation policy instrument in Finland. The Electronics Laboratory of the research centre is a high technology research laboratory which is located far from the industrial centres of Finland.

The location of the Electronic Laboratory has had no negative impact on its development. The laboratory also seems to have been able to contribute to local industrial development.

The Electronics Laboratory has been able to perform significant technology development projects and to transfer the results effectively to the users of the new know-how. The laboratory is strongly directed toward product development on a contractual basis. The main reasons for performing research projects in an outside laboratory seem to be the superior resources and know-how of the laboratory or a shortage of in-house research resources. However, the services of the research laboratory were not considered indispensable.

The industrial customers seemed to use contract research in product development. More responsibility for the whole innovation process was given to the research laboratory than had been expected. This was reflected by the share of the costs of the contract project compared to the costs of the whole innovation project and the fact that the representatives of the companies considered the contract projects important or very important to the innovation process. None of the innovation projects had so far turned out to be a failure, which indicates that the most risky projects are probably performed by the customer companies themselves.

A public research laboratory can be an effective innovation policy tool as the evaluation of VTT's Electronics Laboratory has shown. However, the success is by no means self-evident. The other three VTT laboratories which were evaluated made their important contributions to the innovation system, but their contributions were not as direct as those of the Electronics Laboratory. The success seems to be dependent both on the competent management of the laboratory and the ability of the industry to use the research services without prejudice.

References

Blume, S.S. (1981)
    "The state of the art in science policy research and implications for research", a report to the Swedish Council for Planning and Cooperation of Research, Stockholm, STU.
Freeman, C. (1982)
    The Economics of Industrial Innovations, 2nd edn, London, Frances Pinter.
Goldberg, W. (1981)
    "Explorations into the instrumentality and innovation policies", FE-rapport NR 196, University of Gothenburg, Department of Business Administration.
Lemola, T., R. Miettinen, M. Ollus, E. Ormala and B. Wahlström (1983)
    "The role and position of a public research institution in innovation management", a seminar paper at the IIASA task force meeting on human factors in innovation management, Helsinki, 9-14 October.
OECD (1983)
    Recent Results: selected R&D indicators 1979 to 1983, Paris, OECD.
Ormala, E. (1983)
    "Supporting technology development evaluation with multiple attribute utility analysis and fuzzy decision analysis", WP-83-59, Laxenburg, Austria, International Institute for Applied System Analysis.
Rothwell, R. and W. Zegveld (1981)
    Industrial Innovation and Public Policy: preparing for the 1980s and 1990s, London, Frances Pinter.
Sahal, D. (1981)
    Patterns of Technological Innovation, London, Addison-Werley.

INNOVATION POLICIES IN EAST ASIA
AND SOME IMPLICATIONS FOR WESTERN EUROPE

Ove Granstrand and Jon Sigurdson

World attention is focusing on East Asia where more than a quarter of the world's population lives. The fascination has always been there because of the exotic flavour of the cultures. Today the impressions are mixed with concern for the economic performance of the actors and expectations for the future.

The region shows a great variety of actors. First, Japan is a category by itself, having closed the economic and technological gap *vis-à-vis* the United States and Western Europe in the decades since World War II. Not far behind are the Asian newly industrialized countries (ANICs) such as Hong Kong, Singapore, South Korea and Taiwan (often referred to as the four dragons). Third comes the People's Republic of China which, although at a low level of economic development, nevertheless makes its presence felt because of its size and future gigantic potential. At the fringes of the region is the group of ASEAN countries, and further west another potential economic giant: India. If we include all countries east of Iran, the region contains more than two billion people.

Following from the globalization of technology, it is important to realize that interaction among countries is becoming more and more intense. For the purpose of analysing the high technology challenge in the industrialized countries, it would be useful to look at the global economy consisting of three different layers of international actors. At the lowest level are most of the developing countries at a modest level of economic development. Generally they attempt to utilize and further develop a comparative advantage in labour-intensive production to promote exports to obtain needed foreign exchange in order to pay for imports required, in particular for further modernization. Many of the large populous developing countries, like China and India, are actively pursuing such a policy. In doing so, some of the better organized and/or larger countries, partly because of their size, are successfully capturing the markets for labour- and assembly-intensive industrial products, by increasing their shares in the global trade in such products. This results in an erosion of the comparative advantage enjoyed by the newly industrialized countries (NICs) forcing them to move up the ladder into capital- and knowledge-intensive activities. Thus the NICs are moving into the sanctuaries of the global elite.

At the upper end of the three-level hierarchy are the industrialized countries, including Japan and Sweden. In recent years it has increasingly been noticed that several elements have converged to create rivalry or tensions in the market struggle for high technology projects. First, almost all industrialized countries consider advanced technologies to be essential for their further economic development by staying competitive in international trade. Second, related to this belief, is the hope that revitalization of industry through high technology infusion would, in Western Europe and the United States, make a major contribution to the reduction of the 30 million unemployed at present. Third, the scientific and technological capability of many countries (notably Japan) has risen so that in several fields the post-war supremacy of the United States is seriously challenged.

In between is a layer of newly industrialized countries like Taiwan, Republic of Korea (South Korea) and Mexico, which are in the process of rapid industrial restructuring. Having for a number of years enjoyed a comparative advantage in labour-intensive production, which has been the mainstay of their export-oriented economies, they now see this advantage being eroded by the developments in the less developed countries — the lower level of the global hierarchy. Thus the NICs have generally interpreted their new situation into policies to favour the development of knowledge-intensive industries and the "informatization" of their economies. In fact they want to establish a new comparative advantage at a higher level of economic development, which brings them into more direct competition with the already mature and fully industrialized countries. So the latter can no longer take for granted that they will stay competitive in industrial sectors, where in the past they enjoyed an advantage because of their engineering skills and capability to to develop, design and manufacture complex systems.

So, at the highest level, the industrialized countries are forced to rethink their industrial and technological priorities in order to stay competitive in this world of change. Large countries with abundant domestic resources and limited involvement in global trade (such as the USSR and to a lesser extent the USA) are less affected. Others, such as Japan and Sweden, are more influenced and are forced to react much more quickly to such changes.

More than half of the population in the Asian countries east of Iran are under the age of 21. This is a contributing factor to why these are temporarily very poor countries. In the next two decades these countries are going to be dominated by people who are in their twenties and thirties. More than one billion people are going to enter into the most productive age groups. Furthermore, we can expect that a majority of them will be literate. This will be an increase in the labour force unlike any situation formerly seen. If most of these people are provided with job opportunities fitting their capabilities, we would then witness a tremendous surge in production and living standards over the next two decades.

The developments and policies regarding science, technology and industry, in Japan and the Asian NICs, should be viewed against the scenario of a rapidly developing and prosperous Asia. This paper will first give a general picture of the ANICs and then, since much has been written about Japan, focus on some specific features of Japanese innovation policies.

A note on terminology: "Innovation" is used to mean "the first introduction of a new product, process or system into the ordinary commercial or social activity of a country" (Freeman et al., 1982, p.201). It may be noted that in this definition, "new" is taken to mean new to a specific country rather than new to the entire global market. Innovation policy is normally not an area of coherent policy-making in a country. Rather, policies for stimulating innovation emerge out of policy-making regarding science and technology and industry. Thus "innovation policy" could be defined as being "essentially a fusion of science and technology policy and industrial policy" (Rothwell and Zegveld, 1981, p.1).

Present situation in the ANICs
The ANICs show several similarities regarding their past developments and policies. First they have experienced a rapid development with a remarkable rate of growth. Second, their policies have changed from, in general, short periods of import substitution to export promotion. This has commonly been cited as an important factor behind the successes. Third, they have taken off in their industrialization process roughly at the same time, that is in the early 1960s. This was probably a very good time for developing leading sectors of industry and laying the foundation for continued export-led growth. Fourth, the ANICs have all experienced political

uncertainties which have had a mobilizing effect. Fifth, the textile sector has been a leading sector in the take-off, and the electronics sector is aimed to be a leading sector in the future. However, this is only a rough picture. Cheap labour in the electronics assembly and export of consumer electronics have also been highly important in the past.

Naturally there are many more similarities which could be enumerated, e.g. their colonial heritage, with its relatively high levels of education and capability of absorbing foreign technology. However, there are also some striking dissimilarities. Obviously there are some important ones pertaining to the size of the population, the political situation, natural resources, etc.

Apart from these, the following dissimilarities may be pointed out. First, the size and ownership structure of industry varies greatly. Industry in South Korea and Singapore is dominated by large firms, in Singapore mainly of foreign origin. Hong Kong and Taiwan have a dominance of small, local firms.

Second, although all ANICs classify themselves as free market economies, the degree and form of government intervention varies. It is far from easy to assess the actual amount of elements of a planned economy in the market economies. Table 1 gives the features of formal planning in the ANICs. A rough assessment would be that government intervention is strongest in South Korea, followed by Signapore and then by Taiwan and finally by Hong Kong with her "minimal interference" policy, which has given Hong Kong a reputation as the last bastion of economic *laissez-faire*.

Third, all ANICs emphasize the build-up of an indigenous R&D capability (see Table 2). However, the route to this goal varies. Hong Kong relies on non-intervention and provides no special incentives oriented towards R&D, apart from some preferential treatment to certain types of industries when attracting them to the industrial estate in Tai-po. Singapore mainly relies on attracting R&D operations of foreign multinational corporations through a package of R&D investment incentives, such as tax incentives, selective R&D grants, prolongation of low-tax pioneer status to firms and access to the Kent Ridge Science Park. South Korea builds up a rich structure of public R&D institutes as an intermediate stage to the build-up of company in-house R&D units. Taiwan has a similar philosophy but has not yet determined her main route in this respect. All of the ANICs encourage technology imports through licensing in, cooperative agreements and joint ventures, but apparently South Korea is most active in this respect. Also, and most importantly, the education of qualified scientists and

TABLE 1

Economic and Industrial Planning in the ANICs

	Hong Kong	Singapore	South Korea	Taiwan
Economic plan	No central plan	10 year economic development plan announced in 1979	Fifth 5 year economic and social development plan 1982-1986	First 4 year development plan introduced in 1953 10 year development plan introduced 1980, supplemented by a 4 year economic development plan 1982-1985
Bodies for economic planning	(Industrial Development Board)	Economic Development Board	Economic Planning Board	Council for Economic Planning and Development
S&T plan	No	?	Yes	S&T development program exists
Bodies for S&T planning and coordination	No	No special ministry	Ministry of Science and Technology plus special Presidential conferences	No special ministry. Sectoral planning coordination by specially appointed minister

TABLE 2

Build-up of R&D Capabilities in the ANICs

	Hong Kong	Singapore	South Korea	Taiwan
R&D expenditures as % of GNP	n.a.	0.2% in 1978	0.48% in 1970 0.91% in 1980	0.30% in 1978 0.55% in 1980 0.76% in 1981
Projected R&D expenditures as % of GNP	n.a.	n.a.	2.0% in 1986	1.2% in 1985 2.0% in 1989
Private sector share of total R&D	n.a.	63% in 1978	40% in 1981	56% in 1981

Sources: Official planning documents

engineers (QSEs) is strongly emphasized. There is an overall shortage of QSEs but strong efforts are made to create comparative advantages in the area of skilled labour and engineering manpower. In this connection efforts are also made, especially in South Korea and Taiwan, to reverse the brain drain through repatriation schemes.

Industry Finance Systems

The industry finance systems also clearly differ among the ANICs and Japan and especially between these countries and the USA and UK. The latter difference deserves special notice. Financial systems can be broadly divided into two types — bank-oriented systems and market-oriented systems (see Rybczynski, 1983). Bank-oriented systems are characterized by a large degree of dependence of industrial firms on finance obtained mainly in the form of non-marketable loans from financial intermediaries and, above all, banks. Market-oriented systems are characterized by a large degree of dependence of industrial firms on capital markets. The USA and UK have a market-oriented financial system, much because their industrialization was a result of private entrepreneurship unaided by the state. Japan, South Korea, Taiwan and also countries in continental Europe have by and large bank-oriented systems. Industrialization in these countries has been assisted by the state, in some cases even engendered by the state.

The equity ratios of firms in countries with bank-oriented systems are typically lower than in countries with market-oriented systems. On a macro-level it is difficult, if not impossible, to assess a meaningful correlation between growth rates and equity ratios. However, there is evidence that a high leverage in combination with fiscal and monetary policies and selective government incentives has been an important factor contributing to industrial growth in Japan.

This leverage supported by bank debt has enabled Japanese firms to pursue growth without being confined by retained earnings while lower capital costs due to the tax deductability of interest costs have permitted, indeed encouraged, lower pricing. This has helped companies to grow rapidly and cover interest costs. Since rapid growth means substantial productivity gains, this situation stimulates further price reductions, investment, borrowing, tax benefits and expansion. Hence, the logical Japanese emphasis on investment and market share. (Rapp, 1977, p.43)

The capital markets are underdeveloped in Japan, South Korea and Taiwan, while they are most highly developed in the USA and UK. There has then been a (somewhat self-deceptive) tendency in the USA and UK to view their financial systems as the world's most advanced and efficient. However, the principal function of capital markets is not to finance industry or provide funds needed for technological developments. The important thing is the ability of the whole financial system to supply a necessary flow of funds to industry under favourable terms for the development of technology-based businesses. This fact has not received sufficient attention.

Country differences in the supply of risk capital of debt and equity type give differences not only regarding the cost of financial capital but also regarding the pattern of entrepreneurship. While Japanese and South Korean entrepreneurship is heavily biased towards corporate entrepreneurship — to an increasing extent shifting from imitation based to innovation based — US entrepreneurship may be overly biased towards autonomous entrepreneurship, hampering the necessary concentration of a new technology-based industry in later stages of its development, when economies of scale and scope might accrue. (It has been claimed that the US industry might even lose out in the semiconductor business, partly due to this, see e.g. Okimoto, 1983.) In light of the various comparative advantages of small/new firms versus large/old firms in various stages of innovation and the maturing of new technologies, country differences in supply of risk capital of different types and cost thus have a direct bearing on the innovativeness and international competitiveness of the industry of a country.

Now, as pointed out by Rybczynski (1983), there is a trend towards convergence of bank-oriented and market-oriented financial systems. The financial systems in Japan, South Korea and Taiwan are under pressure to be liberalized and financial markets will be opened up for a number of reasons. For example, Japan is in the early 1980s attempting to create a venture capital market (in fact the second attempt), raise equity ratios and facilitate autonomous entrepreneurship. (However, in contrast to the United States, Japanese venture capital firms are created by large banks, securities companies and trading houses.) There is an expressed need that Japanese firms become more concerned about their capital structure, equity ratio, and cost of capital for a number for reasons, among other things because of their increasing internationalization in terms of foreign direct investment (Kurosawa, 1981). Naturally there are many factors behind the present and approaching changes in the financial system in Japan. One factor is the pressure on Japan exerted by the USA, another is the entry of Japanese industry into a truly innovative rather than an imitative stage, both of which put new requirements on financing.

Indications of a transitional period

A methodological note is first in order. It is always difficult to assess whether an economic object such as a country at the time of observation is in a certain stage or in transition between two stages. Also it is doubtful if such an assessment is best made by an inside or an outside observer. (Certainly a first-time observer is at a disadvantage.) Moreover, the assessment of stages and transitions between these is not only a matter of observability and observation errors but also a matter of what criteria to use in defining certain stages and transitions. Generally speaking, transitions are characterized by rapid and transient changes in key variables. Factors inducing such changes may be classified into endogenous and exogenous. This distinction is not without its problems. First one has to determine a borderline between the object under study and its environment. Second, different factors may interact so intimately that it is difficult to separate them into different categories. Nevertheless, as is the case here, when different countries are more or less simultaneously in transition, it is normally useful to look for common exogenous factors.

Indications of a transition are recent changes in technology and industrial policies in the ANICs. In general, there is a new emphasis in these on higher value-added through moving into high technology products. Thus, for example, Singapore announced in 1979 a ten-year economic development plan, emphasizing the movement from low-wage, labour-intensive industries like textiles and electronics assembly into more capital- and skill-intensive industries. A main instrument for restructuring the economy along such lines was an across the board increase in wages, higher than the growth in productivity. Other instruments have also been implemented, such as R&D tax incentives and the creation of the Kent Ridge Science Park.

Hong Kong has also in recent years announced her wish to move in similar directions. However, no strong government instruments have yet been developed for this purpose.

Taiwan speaks of her fifth stage of industrial development, starting in 1981, as the stage in which certain technology-intensive strategic industries will be developed.

South Korea considers that the successful execution of her Fifth Five-Year Economic and Social Development Plan for 1982-1986 will push her ahead to an advanced industrial state.

In order to reach the stage of the advanced industrial societies technology transfer from abroad will be accelerated and domestic capacity to develop technology will be cultivated. Consequently, Korea's technological level will be raised in the semi-conductor, computer and systems industries. (Fifth Five-Year Economic and Social Development Plan, p.121)

The build-up of an indigenous R&D capability in the 1980s is commonly emphasized by the ANICs, although less so in Hong Kong where R&D is still a foreign concept.

Another common feature among the ANICS is the emphasis on the electronics sector, especially industrial electronics, and the de-emphasis on the textile sector.

If recent policy changes towards upgrading the technology factor are taken as an indication of a transitional period, what factors then account for these changes? Since the changes by and large are simultaneous in the ANICs it is tempting to start to look for explanatory factors in their common environment. First the ANICs have found themselves caught in a kind of sandwich position. Their traditional comparative advantages of cheap labour have eroded. They are pushed from below by the near-NICs of Indonesia, the Philippines, etc. which out-compete them in their traditional labour-intensive industries. At the same time their high technology sectors are not yet sufficiently developed to compete with industrialized countries (ICs) such as Japan and the USA. Second, they are all heavily dependent upon foreign trade. The past successes of ANICs and Japan on IC markets have created protectionist reactions in the ICs. This in turn has further induced the ANICs and Japan to move into high technology products, which are considered less sensitive to protectionistic measures. Third, the global recession has reinforced both the forces of international competition and the protectionist reactions. Fourth, Schumpeterian waves of investment opportunities in new technologies may be considered to open up possibilities for new entrants also from non-ICs. It is well known that in many sub-sectors of the electronics sector the barriers to entry are low, since, for example, patent protection, static economies of scale and customer loyalty are less prevalent. Fifth, new automation technologies and wage changes have changed the pattern of foreign direct investments and those countries which relied on attracting foreign direct investments for assembly operations (for example Singapore) must look for other types of employment. Finally, there is apparently also a bandwagon effect among the ANICs. This is especially strong between Hong Kong and Singapore and between South Korea and Taiwan.

Barriers from past development and policies in the ANICs
Now, what kind of barriers are there to the ANICs moving into a

stage of innovation and R&D based development? Naturally there are all kinds of problems in making such a transition. Not all of these problems derive from past developments and policies but a few important ones do. Mostly these appear to be country specific. The successful reliance of Hong Kong on non-interventionism has certainly created a barrier to actively promoting the build-up of an indigenous R&D capability. It is not likely that such a build-up will take place with sufficient pace if left entirely to incentives through the market. This is especially so given the tradition of trade and banking, which gives a short-term profit orientation, often with a conservative bias, in investments. The flexibility of Hong Kong entrepreneurs in imitative manufacturing and trade of mainly consumer products is widely acknowledged, but it takes quite another breed of entrepreneur to start up high technology based ventures. Besides, there is a shortage of long-term venture capital in Hong Kong. The whole culture of imitative manufacturing, trade, conventional banking and *laissez-faire* probably has to be drastically altered if a high technology based industry is going to emerge. New generations may be instrumental in such a cultural transition and in fact there are prospects that the children of the traditional entrepreneurial families will form a new generation of foreign trained, high-tech entrepreneurs. Whether this will materialize is still in the air and certainly government policies will play an important role. It is our conjecture that the Hong Kong government will become more technology promoting and interventionist in the near future. A major source of uncertainty, however, is the future relationship with the People's Republic of China.

Singapore, with its limited land, natural resources and labour, found the multinational corporations (MNCs) to be a fast vehicle to industrialization. The build-up of an indigenous manufacturing industry thereby suffered, probably to an irreversible degree, at least in a medium term perspective. The past policy of relying on MNCs probably went too far and it might very well be too late in the 1980s to create a viable indigenous manufacturing industry of reasonable size. The entrepreneurial base is simply too small and cannot easily be developed. The MNCs absorb the engineering and managerial talents, and incentives to technology start-ups outside the MNCs may be too weak under conditions of full employment, high wages and an atmosphere of "cosyness". Besides, the sources of new technology would rather be the MNCs in order to create such spin-offs. It is of course conceivable that as an intermediate stage the MNCs could be stimulated to locate R&D units in Singapore and could function as sources of new technology based firms as a basis for building an indigenous industry. However, another and perhaps better start-up process is sending talented students abroad for study and junior employment for some years in foreign firms abroad, and then attract them to repatriate and start new firms. This is in contrast to attracting foreign firms to locate in the country and then stimulating spin-offs from there. In any case, it is not an easy thing to do.

It is conceivable that government authorities in Singapore have already realized the strong barriers to building an indigenous industry in the 1980s and have chosen to continue to rely primarily on MNCs with ancillary industries. There are special problems in doing so when moving into high technology areas, but by and large it may be an economically feasible and preferable route in total, since the service industries may become more important in the long run in the Singaporean economy. These industries are largely indigenous in Singapore and some of them may be technology-intensive as well. In a future long-run transition into a service economy, the present stage with MNC-dominated manufacturing may turn out to have been just an intermediate stage, in fact shortened by the choice of MNCs as a fast vehicle in the transition.

South Korea has developed along a path of rapid but unbalanced growth. Several of the past developments and policies present barriers to a transition into an innovative stage. First there is a strong dominance of large firms in the economy and small and medium sized firms are few and badly developed. Although it is a common argument that large firms tend to stifle innovations, this does not necessarily apply to Korean large firms since these are relatively young, say 35-40 years at most, and barriers to innovation in large companies often derive more from the old age of the companies than from the size of them. More serious than the sheer size of the companies are the two internal barriers to promoting technologists in large South Korean companies. As is often the case in Chinese-owned industries as well, there is seldom a separation between ownership and management, in contrast to what has developed in Western companies during the last century. Efforts have been made in South Korea and Taiwan to float off the big family-owned firms on to the stock market in order to, among other things, improve the quality of top management, but without much success. A second promotional barrier is that engineers are recruited at lower levels in the organization and then are not in general promoted even to middle management positions. This barrier is considered even more difficult to overcome than the first one. It has two effects. There will be no engineering competence in middle and top management positions and there will be many old engineers in engineering positions in the future. Both effects clearly may work as barriers to innovation. As in Singapore, the entrepreneurial base in South Korea has been underdeveloped due to the dominance of large firms.

Second, if the degree of government interventionism probably is too low in Hong Kong for entering an innovative stage, it is probably too high in South Korea. There are strong arguments for decentralization in promoting creativity and innovation.

Third, the commitments to heavy industries have been too strong. However, this is partly due to military reasons.

Fourth, a major issue in the early 1980s concerns the proper pace at which import liberalization ought to take place. However, it is difficult to assess whether past import substitution policies have built-in barriers to transition. Certainly it may be argued that technology based infant industries need a period of initial protection. The important point is that during such a period domestic competition must be at least temporarily stimulated. In contrast to Japan, the degree of domestic competition during protection from international competition has been very low in South Korea.

Taiwan, finally, has in some sense been oversuccessful. The entrepreneurial base is perhaps too broad, which has led to a too fragmented industry. Government efforts to stimulate necessary mergers, e.g. in the electronics sector, have been less than successful. Moreover, good profits in the past years have not given any stimulus to allocate resources to in-house R&D in the companies. The industrial R&D is presently found mainly in large firms, which are relatively few in number. The export successes have taken place largely without developing indigenous skills in international marketing. Approximately 60% of exports from Taiwan are managed by Japanese trading companies. As already mentioned, there is also the problem with the frequent low quality of top management in several old family-owned companies. Moreover, there is now a severe shortage of qualified scientists and engineers, a shortage which is aggravated by a brain-drain as well as by rapid transitions of engineers into management positions (in contrast to the case in South Korea). This is in line with the Chinese culture in which managerial activities are more esteemed than engineering activities. The textile sector has been responsible for much of the success of Taiwan and is still viable, but is now consuming resources that could probably give a better rate of return in the long run if employed elsewhere in the economy. Thus it is difficult to name the textile industry as a declining sector without creating much controversy. In this regard the problem is one of exit barriers in industrial restructuring in general.

Innovation policies with particular reference to information technologies in Japan

Japan's competitive edge was in the past — in the period immediately after World War II — in labour-intensive products, exemplified by textiles. The country's comparative advantage shifted to capital-intensive industries such as steel, petrochemicals and shipbuilding, the latter being labour- (or rather skill-) intensive as well. More recently Japan has emerged as a dominant force in the global manufacture and distribution of automobiles. Simultaneously, the country has been able to capture in succession large shares of the global markets for radios, black and white TV sets, colour TVs, electronic watches and video tape recorders. Most of these achievements have benefitted from government macro-policies with regard to finance and import policies, in combination with company strategies. No doubt the governnment has shouldered the important role of providing the necessary educational facilities to match the industrial and technological development. Government laboratories have been established and expanded. In shifting the sectorial focus in the economy, the ministries concerned have exercised a control and influence on technology transfer and protection of infant industries. However, this influence cannot be seen as primarily aiming at industrial innovations. This situation changed during the 1970s and we will illustrate the new situation with a description and analysis of the engineering research associations (ERAs) which have emerged as a significant policy instrument, particularly in the information technologies.

The ERAs, which are established under the guidance of the Ministry of International Trade and Industry (MITI), have the objective of organizing the major industries concerned to solve technical problems common to the sector or a part thereof. The initiative for establishing an ERA often comes from MITI, basing itself on ideas from their government laboratories, university professors and naturally opinions voiced within the industries concerned. On the basis of carefully worked out reports and plans, an ERA is established as a legal entity made up of the companies who were keen to join or coerced by MITI to become members. The formula for government financial support may differ but in some recent high technology projects — such as very large scale integration (VLSI) — the government has provided grants matching the contributions made by the private companies. Thus the government has over the period from 1961 to 1983 contributed in the region of 500-1000 million million yen to the ERAs. However, the government's impact on industrial research within the private companies may be much more important in its indirect catalytic effect, although this is difficult to quantify.

The starting point for the engineering research associations was a study tour to the UK in 1953 to investigate the British research associations combined with a study of jet engines manufacture. A subsequent report, discussions and articles in technical journals persuaded the Ministry of International Trade and Industry that it would be desirable to introduce this policy instrument into Japan. The research associations were seen by MITI as a powerful tool to increase the country's competitive power in certain industrial sectors. The ERAs have the following benefits according to a MITI publication (1982):

1. Risk-sharing and cost-sharing among participating units.
2. Pooling of resources to speed up the research process and eliminate overlap.
3. A comprehensive research approach, which means that resources are pooled horizontally among enterprises and also vertically through to marketable products.
4. Exchange of information which raises the technological level through the relevant system of an industrial sector.

The ERAs are non-profitmaking organizations which are established according to certain rules contained in a law which was promulgated in 1961, when the first ERA — on high polymers — was established. Since then, up to the end of 1983, altogether 71 ERAs have been established involving more than 500 companies.

In the late 1950s, before formalizing the ERAs in 1961, MITI had experimented with the concept by applying it to the car components manufacturers, where Japan at the time suffered from great deficiences. The apparent success then prompted the formalization of the ERAs which, however, differ distinctly from the British approach. In the UK the research associations have mainly been used to support small companies which cannot climb the necessary R&D threshold on their own. Furthermore, the associations are generally found in mature industries. The Japanese approach was already in its early stages quite different, and has become more so in recent years. First, only a few of the ERAs established during

1961-65 (after which there was a lapse until 1971) have a large number of participants. The largest one was the optical technology research association — 44 members — which provided the technological basis for continuous improvement for the camera manufacturers and is seen as an outstanding success. The average membership was about 12 companies per ERA since 1971, against 17 in the 1960s.

Second, none of the ERAs is a permanent organization. The average length of existence was 11.5 years for the 12 ERAs which were established during the 1960s. Using similar statistics, available only for those which have been terminated, it appears that the length of existence has been drastically shortened to around six years. Third, major companies constitute the dominant force within the ERAs. In the early period there was a substantial membership of smaller companies, but this is no longer the case — with a few exceptions. Thus, the target of an ERA is clearly on technology development as such rather than assisting small and medium companies, as is the case in the UK and most other industrialized countries which are using the same policy instrument.

The role of the major companies becomes apparent from the fact that some 30 of the major companies account for altogether 258 memberships out of a total of 878 in the 71 ERAs. These companies fall into eight industrial sectors: fibres, chemicals, glass, steel, non-ferrous metals, machinery, shipbuilding and electrical machinery. The last could more appropriately be called information technology industries, and will be discussed in some detail. In all sectors with the exception of electrical machinery industries, the ERAs have in the main been geared to singular technology projects which have served the diversification interests of the company. In most of the eight sectors, with the exception of steel and again electrical machinery, it is possible to identify a continuity of interest among the participating companies. Five major steel companies have participated jointly in seven ERAs and it can be assumed that a network has been established through this process. But most of these ERA projects basically are serving the diversification interests of the companies, which of course is a sound indication of the restructuring of the Japanese economy.

It is only in the electrical machinery industry that a pattern has emerged where a limited number of major companies participate in ERA after ERA, where the technology focus has been on areas which are in the mainstream of development within the companies. These companies are Hitachi, Toshiba, Misubishi Electric, NEC, Oki and Fujitsu and the technologies are computers and semiconductor components. The involvement in the various ERAs is listed in Table 3, which also lists two projects which are not formally organized as an engineering research association, although the pattern of company membership is very similar. Recently a discussion has started on how to develop technology for the 100 megabit integrated circuits, which might indicate a second VLSI project.

Today there is a consensus that the VLSI Semiconductor Project in Japan, established in 1976 as an engineering research association (*gijutsu kenkyu kumiai*), was an exceptional success in promoting technological development. The focus of the project was clearly on microfabrication technology, which in essence meant improving and developing new lithography methods. Crystal quality and an improved understanding of mechanical, physical and electrical properties was a second focus. The project resulted in firmly raising the level of VLSI manufacturing technology of the five participating companies: Toshiba, Hitachi, NEC, Fujitsu and Mitsubishi Electric. This is a major contributing factor for the increasing shares taken by these companies of the global market for memory circuits.

Furthermore, the project similarly raised the technical level of two major groups of supporting companies. First, the silicon vendors, although not formally participating in the project, were drawn into a close interaction with the project, which rapidly increased the understanding of how silicon crystals could be considerably improved, at a time when the global as well as the domestic demand was rapidly increasing. Second, a substantial number of equipment manufacturers, companies attached to the five participants as well as independent companies, in a similar fashion considerably improved their understanding of what types of equipment would be technically and economically most suitable for the VLSI circuits to be manufactured a few years hence. So, not only have Japanese VLSI companies established themselves among the world leaders, but crystal and VLSI equipment manufacturers in many fields have established themselves in the top league.

We will now look more deeply into the conditions which made this success possible. There can be no doubt that the high level of funding was a very significant factor. It meant that the total VLSI project research funds between 1976 and 1980 constituted approximately 23% of total sales of Japanese semiconductors in 1975. However, Sakakibara (1983, p.32) mentions that it has been estimated that between a quarter and a third of the total project funds were spent in the United States to purchase the most

TABLE 3

Major Japanese Electrical Machinery Companies having Participated in MITI-sponsored Information Technology Development Projects from 1962 and Onwards

ERA No	7	16	17	18	22-26	33	36	42	47	53			(100
Project	LSC	SC	NC	SC+	SOFT	VLSI	PIPS	NGT	OPTO	HSC	ICOT	3-IC	Mbit)
Start	1962	1972	1972	1972	1974	1976	1977	1979	1981	1981	1981-	1981	(pro-
End	1972		-76			-80	-81		-86	-89		1991	posed)
Grant $billion			52			35	26		19	24			
6501 Hitachi		x				x	x	x	x	x	x		
6502 Toshiba			x			x	x	x	x	x	x	x	
6503 Mitsubishi El.				x		x	x	x	x	x	x	x	
6701 NEC	x		x			x	x	x	x	x	x	x	
6702 Fujitsu	x	x				x	x	x	x	x	x		
6703 Oki	x			x				x	x	x	x	x	
6752 Matsushita El.									x			x	
6753 Sharp								x				x	
6764 Sanyo El.												x	
6781 Matsushita								x					
NTIS NEC/Toshiba			x			x		x					
CDL Hitachi/Mitsubishi/Fujitsu						x		x					
Other L	-	1	-	-	2	-	-	-	6	-			
Other UnL	-	2	4	1	34	-	-	-	2	-			
Total	3	5	7	3	36	7	5	10	15	6			

Explanations to Table 3:

7 LSC: Electronic Calculator (Computer) Engineering Research Association (ERA)

16 SC: Very High Capacity Computer Development ERA

17 NC: New Computer Series ERA

18 SC+: Very High Capacity Electronic Calculator (Computer) ERA

22-26 SOFT: "Software development"
- management planning ERA
- administrative processing ERA
- design and computation ERA
- operations research ERA
- automatic control ERA

33 VLSI: VLSI ERA

36 PIPS: Pattern Information Processing System ERA

42 NGC: Computer Basic Technology ERA

47 OPTO: Optical Applications System ERA

53 HSC: High Speed Calculating System for Science and Technology ERA

ICOT: "Fifth Generation Computer" Project

100 Mbit: Future Ultra VLSI Project (proposed to MITI March 1984 but not yet decided)

Other L: Number of other participating members listed on the stock exchange

Other UnL: Remaining ERA members not listed on the stock exchange

Total: Total number of participants in each ERA

The four digit code above the company name is the "Securities Identification Code".

sophisticated equipment available at the time. Sakakibara also mentions a number of other favourable conditions which were extrinsic to the VLSI project. First, everyone concerned knew that the project was to counter the possible threat of IBM developing a super-chip which — if true and successful — would most likely erode the basis of the computer industry in Japan. This awareness was in the words of Sakakibara "favorable for the integration, motivation, and concentration of research efforts of many members" — in spite of the fact that this assertion was already found to be incorrect in the early planning phase.

A second favourable condition was that the five participating electrical machinery companies had already participated in joint projects under MITI guidance and had already accumulated considerable experience in how to handle matters such as personnel, budget, patents, etc. Thirdly, the timing was also good from a technological point of view, as the time had come to make the transition from optical lithography to electron beam lithography and make preparations for future use of x-ray lithography. At the same time there existed in Japan a substantial scientific basis for high-quality development work on crystal technology and microscopy, which made it possible to make a leap forward.

Related to the existence of a technological threshold to be climbed, the time had also come to switch from individual research to group-oriented research because of the complexity and interdisciplinary character of many of the problems. This in turn favoured the idea of establishing a joint laboratory, which may have been a major instrument in integrating the various development capabilities of the participating companies.

More recently the idea of joint laboratories has been incorporated in several of the MITI-sponsored projects aiming at the development of information technologies. This is a case for the Optoelectronics Engineering Research Association started in 1981 which, with six laboratories and almost exactly the same membership as in the VLSI group, appears to be a true copy. A similar approach has been followed for the Fifth Generation Computer Project, initiated a couple of years ago. A similar approach has also been followed for the R&D Institute of Metals and Composites for Future Industries, although not formally organized as an engineering research association. All these institutes have been established for a certain duration, which is not likely to be extended, with the implicit expectation that they will repeat the success of the VLSI project.

The idea of having a joint laboratory of a fixed duration did not come naturally to the participants in the VLSI project. However, NEC and Toshiba had been cooperating in a joint laboratory, NTIS, since 1972. A third laboratory, CDL, was operated by Fujitsu, Mitsubishi Electric and Hitachi. In fact, the VLSI *kenkyu kumiai* was formally made up of seven participants and two of the units were the CDL and NTIS laboratories. Sakakibara mentions that there was no precedent of a *kenkyu kumiai* having a joint laboratory and furthermore having research members coming from competing companies.

Sakakibara makes the point that "it was neither MITI nor the five private companies, but the ruling Liberal Democratic Party (LDP), that initiated the idea of setting up a cooperative laboratory in 1975". This was prompted by fear that the computer manufacturers in Japan could not stand up to IBM. Apparently the manufacturers were first opposed to such an idea and were much more in favour of government policies or subsidies that would allow them to be on their own. The political interest in supporting the computer industry dates from a much earlier period. Similarly to the situation in the car industry, the attempts to restructure the computer sector had not been very successful. The new approach tried in 1975 was to influence and support the computer industry more indirectly — through technology support programmes. Thus, under political pressure, according to Sakakibara, the major companies were forced to accept two joint laboratories among themselves which also formed the basis for a much wider project : the VLSI Kenkyu Kumiai, with its joint laboratory.

Sakakibara further argues that the VLSI benefitted greatly from factors which were intrinsic to the VLSI project, and in particular he stresses communication and institutionalization. First, the joint laboratories, consisting of the planning section and six laboratories, had laboratory heads drawn from each of the five companies and the Electrotechnical Laboratory (ETL), which is directly controlled by MITI. All the six laboratory teams included members from several companies. It had been indicated to the companies which members would be desirable to have in the project, but it was naturally left to the companies to make the final decision. Then it was left to the teams so selected and the research director, Professor Tarui from ETL to work out the research themes in considerable detail. This was a very lengthy process. However, the time may have been well spent as it narrowed down a possibly unwieldy project and most likely achieved a consensus among the researchers. This is likely to have formed a very strong basis for a continued interchange of ideas among the researchers throughout the project period.

On the formal side there are two management aspects which are important. The project leader was a retired MITI official, Mr Nebashi, who has since joined IBM Japan, while the research director came from ETL. The two leaders, according to Sakakibara, were different in their abilities, characters and personalities and undertook to carry out a leadership role which fitted their personalities with a clearer understanding of a subtle dividing line for their respective responsibilities. As a result, the two leaders actually made the organization into an institution which became very efficient in achieving the objective for which it was established. So, aside from the extrinsic factors already mentioned, the VLSI project greatly benefitted from the institutionalization of the formal organization and the high and efficient level of communication within it.

The success of the VLSI project is, according to Imai (1983), a Japanese economist, one more indication that future industries will be characterized by the accumulation of know-how related to industrial technology, which is built up not only by the company which uses that technology to produce products. What is equally or even more important is the process of accumulating the required industrial technology through feedback between manufacturers and the makers of machinery, which incorporates the technology into equipment and devices.

The approach has characterized the development of industrial technology in Japan in the past couple of decades. Imai points out that production plants in Japan do not rely on a specific technology but a system under which production technology is gradually improved in response to requests by users and in cooperation with machinery makers. Therefore a company anxious to promote technical innovations can regard its organization for achieving such an objective as a large system which reaches far beyond its corporate walls and includes related machinery makers and users.

In the past it has been taken for granted that the size of the company is a major factor when considering the efficiency of "large scale" technologies. The comments above would, however, indicate that (in the words of Imai):

the factor determining the effect of technical innovation is the size of the "system" related to the process of acquiring knowledge for the development. At the same time, the efficiency depends on how well know-how and information are transmitted and transferred.

Imai goes on to say that if Japanese companies are achieving good results in the new phase of technical innovations, it is because of the scale and economy of their system and the efficiency with which know-how is transmitted and transferred.

Backward integration into joint research: an implication for Western Europe?

In discussing science and technology in the industrial field, it is necessary to consider four characteristics which have become more important in the last couple of decades. First, many technologies have become global in nature, requiring huge markets and huge R&D resources. Second, there has been a merging of interst among multinational companies, governments and other interest groups, which at least until recently also included labour (trade unions) and consumers. Third, the development of new technologies is based on complex interaction among researchers in companies, universities and laboratories which form large networks. Fourth, joint research — still mainly of a national character — is used to speed up development, reduce costs and risks among participants.

A high level of technical and scientific competence constitutes a national advantage, which can be compared with wealth in capital or natural resources. This advantage lies primarily in "human capital" — the skills and knowledge of the work force — and in a variety of institutional research facilities and traditions which may be very important. Advances in certain technological fields can often act to substitute for inputs of natural resources and physical capital. However, technical and scientific knowledge is the one of the three factors which can most easily be eroded.

Technology is mobile and transferable when the recipient country has sufficient skills and knowledge. Thus the position of the United States and Western Europe must not be narrowly viewed with regard to changes in the rest of the world. The policies for R&D, education and infrastructure development in several of the NICs — in particular those discussed earlier — and in some of the big developing countries like China, pose additional threats of undermining the comparative advantage enjoyed by the USA and Western Europe. This has already happened in consumer electronics, where Japan, followed by the NICs in the region, has come to dominate not only global trade but also the domestic markets in the United States and in many of the Western European countries. The major European countries are already at a disadvantage in several sectors of information technology, such as computers and integrated circuits, where even the USA feels threatened by the recent advances of Japan.

It is against this background that one should view the initiatives taken by certain segments of the high technology sectors in the United States. So far, the initiatives within the electronics sector have received most of the attention because of the size of the projects and the outspoken ambition to tackle what is perceived to be a Japanese threat, which might be followed by a South Korean threat. Some projects will be briefly mentioned.

The Microelectronics and Computer Technology Corporation (MCC)

This organization was established in 1982 after the Chairman of Control Data Corporation convened a meeting of the top executives in major US computer and semiconductor companies. The organization will basically follow the model of the engineering research associations in Japan. This means that the participating companies will provide scientists and researchers to MCC for a period of up to four years. Furthermore, the participating companies will also fund the activities of the organization. The consortium was at an early stage joined by twelve companies, but key companies such as Cray Research, Texas Instruments and Intel, as well as IBM, decided to stay out. It was reported that the consortium would have a budget of $75 million a year and a staff of 250. The member companies join in various projects/programmes on an individual basis. The planned projects also include an ambitious ten-year programme aimed at breakthroughs in computer architecture, software and artificial intelligence. There are great similarities with the Fifth Generation Computer Project in Japan and MCC has been created explicitly as a US response to the successful cooperation of Japanese industry and government in advanced electronics.

The Semiconductor Research Cooperation (SRC)

This organization has also been established with the expressed purpose of countering the Japanese threat to the dominance of US companies in the markets of semiconductor devices. Japan has already captured an important segment of the market: the so-called random access memory chips (RAMs). Today Japan supplies almost 70% of the 64K RAMs and it appears likely that Japanese companies will also capture a large share of the next generation of memory chips: the 256K RAMs. The same companies may even be able to move into the microprocessor markets where the US companies still dominate.

The SRC is made up of 13 manufacturers of integrated circuits, including Control Data, Digital Equipment, Hewlett Packard, Intel, IBM and Motorola. At the time of initiation it was reported that the annual budget would increase to a level of $30 million in 1984. In contrast to MCC, the SRC does not carry out its own research. Instead it sponsors research at the universities and the establishment of centres of excellence in various scientific disciplines and provides project awards. The research cooperative also addresses specific front-line technologies such as three-dimensional chip structures and new manufacturing technology.

The two organizations should be seen as complementary efforts in re-establishing the US superiority in integrated circuits and computers or at least preventing the further erosion of the US competitive situation. It is still to early to judge the US way of meeting the Japanese challenge by forming research consortia. However, it is clear that defence programmes have in the past been instrumental in giving the United States the technological lead in many sectors. Because of national security concern for leadership in frontier technologies, it is natural to expect that defence will take the lead in certain applications of semiconductor devices and computers.

If the US government and companies are concerned about the competitive situation in electronics, there may be a still stronger reason for the European companies and governments to take remedial action. Europe may have been able to maintain a sizeable foothold in consumer electronics such as colour TV and radios, compared with the United States, but its position in computer semiconductor devices is constantly, in relative terms, deteriorating. Since the late 1960s, the European Community (EC) has unsuccessfully attempted to establish a common science and technology policy. Such a policy, it has often been argued, would contribute to the vitality of the participating countries and to the community as such. On an individual basis there are both successful and unsuccessful cases, exemplified by the Airbus Consortium, which until recently has captured a sizeable share of the global market for wide-bodied jet aircraft, and Concorde, which has been an appalling failure.

However, today one can see the US approach in meeting the Japanese challenge and similar attempts have occurred almost simultaneously both within the EC and in individual countries. In late 1982 the EC and a dozen electronic companies established a small venture under the name of European Strategic Program for Research and Development in Information Technologies —

ESPRIT. The European Commission in Brussels prepared a draft proposal based on the initial concept, which calls for joint allocation of \$1.3 billion over a period of five years, which is to be jointly shared by EC and industry. The reported aim is for the European information technology industry to become competitive within the present decade. The following five sectors have been singled out for development: advanced microelectronics, software, data processing, office automation and the application of computers in industrial manufacturing. The proposal has encountered a number of difficulties, which have had less to do with the substance and form of cooperation but rather the different views on the EC budget, where individual countries have delayed the start-up. In the meantime it appears that the European countries have decided to concentrate on national programmes which are worth while, but are not likely to match the government-industry efforts in Japan.

The emergence of research consortia in the electronics sector in Japan indicates a need to find ways of internalizing the research element as opposed to development. Internalizing backward into raw materials or forward into distribution is sometimes natural and economically efficient for enterprises in many industrial sectors. Integrating or internalizing backward means integrating into production of raw materials for enterprises in traditional manufactures. It means internalizing research for knowledge-intensive companies in sectors like telecommunications, electrical equipment, pharmaceuticals and transportation. The backward integration into research will take different shapes depending on national structures of research laboratories and universities. The enterprises are likely to greatly influence the flow of research resources, directly or indirectly, from demands spelled out by the market in order to meet societal and industrial needs. Thus one can expect a growing role for enterprises in coordinating and integrating research in the form of research consortia. A number of propositions follow from an emerging new situation:

1. The modern industrial enterprise accepts joint research, preceding product development, when (government) coordination clearly provides lower costs and reduced risks for subsequent product development — a situation which requires expanding or captive markets.
2. Joint research programmes do not come into existence unless the interests of the state and the enterprises coincide. This happens naturally within the military sector or in states which have shouldered a development interventionist role and are able to

maintain a high rate of growth. Global scale cooperation may come into existence without state intervention.

3. Advantages of internalizing the research stage jointly with other enterprises do not arise until the speed of technical change and magnitude of required resources post problems which are more than offset by risks associated by research jointly with competitors.

4. Joint research favours the creation of large-scale projects requiring resources for a considerable time — longer than the usual planning horizon in private companies.

5. Large enterprises involved in joint or cooperative research of a long-term nature are increasingly going to dominate the industrial structure.

References

Freeman, C. and Soete, L. (1982)
    Unemployment and Technical Innovation. A Study of Long Waves and Economic Development, Frances Pinter, London.
Imai, K. (1983)
    "Japan's Industrial Society: Technical Innovation and Formation of a Network Society", *Journal of Japanese Trade & Industry*, no.4, pp.43-48.
Kurosawa, Y. (1981)
    Corporate Financing in Capital Markets, Research Institute of Capital Formation, The Japan Development Bank, Tokyo.
Okimoto, D.I. (1983)
    "Pioneer and pursuer: the role of the state in the evolution of the Japanese and American semiconductor industries", Stanford University, California.
Rapp, W.V. (1977)
    "Japan: its industrial policies and corporate behavior", *Columbia Journal of World Business*, Spring, pp.38-48.
Rothwell, R. and Zegveld, W. (1981)
    Industrial Innovation and Public Policy: Preparing for the 1980s and 1990s, Frances Pinter, London.
Rybczynski, T.M. (1983)
    "Industrial finance system in Europe, US and Japan", paper presented to the IUI conference on the Dynamics of Decentralized (Market) Economies, Stockholm-Saltsjöbaden, August 28—September 1.
Sakakibara, K. (1983)
    From Imitation to Innovation: the Very Large Scale Integrated (VLSI) Semiconductor Project in Japan, Alfred P. Sloan School of Management, WP 1490-83, October.

THE APPROACH OF DESIGN AND
CONCEPTS OF INNOVATION POLICY

Jean-Eric Aubert

There are two parts to this paper. The first presents design, in its very nature, as the art of product creation, and highlights its importance in any innovation process. The second part discusses the promotion of design within innovation policies, as it has been approached in the past and as it could be in the future.

The art of product conception

Styling, functional aesthetics, packaging — these words are often associated with design. For a large part of the whole community, including a number of industrialists, the word "design" evokes cheap products with fashionable aspects. However, as a matter of fact, design is something quite different. It is the art of product conception itself.

Design starts with the definition of the product's objective. Then, concerned with the whole coherence of the product to meet its objective, the task of design is to integrate all the determining parameters: nature and characteristics of materials and components, ergonomics, quality, price etc. This requires information queries, feasibility studies ... up to the obtaining of prototypes ready for production. Design does not confine itself to the product, but it may also concern the related production process (engineering design).

When industry needs to come back to the fundamentals of product conception, it (re)discovers design. A good illustration is given by the drug industry. A recent article in *The Economist*, "The drug industry moves from discovery to design" (7 January 1984, pp.71-74), makes the point clear. Pharmaceutical research has been based on methods of molecular screening; billions of dollars have been spent on hazardous testing to discover "active ingredients" with very a low probability of success, decreasing with time.

Several factors suggest that chance and hard slog will play a smaller role in the search for new drugs. One is a limiting factor: the sheer cost of getting a new drug through the R&D stage and the market argues against trusting to chance to ensure an adequate success rate. The other two are enabling factors: new knowledge of how the body works and new technology are now making rational drug design look a practical proposition.

The principles of drug design were in fact set up almost a century ago by the German Paul Erlich, who introduced the notion of drug specificity or drug targeting when he discovered that certain dyes were taken up by parasites and killed them off, while leaving patients' tissues unharmed.

This is an important example. It shows that design, as a basic methodology, is relevant to the most science-based industries. It cannot be confined, as it tends to be, to "marginal" industries concerned with aesthetic factors, such as furniture or building. As a global targeted approach, design is a basic methodology, a kind of *état d'esprit* which determines, pervades and controls the whole innovation process. Design appears as a kind of discipline of industrial creation.

In doing this, design fulfils a critical function at several levels. It questions the practice and efficiency of research (applied and technical). It also questions the whole management of the firm, its internal organization, communications and compartmentalization of different services. It also leads to reassessment of the relationships of industry with its whole environment: users, suppliers, competitors.

The reasons why design has been misunderstood and indeed marginalized are complex. However the pervasive development of the ideology of scientism obviously has much to do with it. Innovation has been presented as the natural result, the end product, of a somewhat mechanistic process, starting with research (production of ideas), following through with development (production of prototypes), and finishing with manufacturing and commercialization (production of goods and services). In such a very broadly adopted view, there is no place for recognizing the key importance of product conception itself.

Meanwhile, partly because of this situation, designers have often isolated themselves from the industrial process. Behaving like a corporation, they have created their own schools and presented themselves as artists rather than engineers. This tendency has been apparent throughout the history of design. It is still alive today: the corporation often presents its members as the key men in the firm, the only ones able to speak and translate the different languages of the specialized services, as the mediators between the industry and the users.

sythesis of images, movies based on computer programming (such as *Tron*) etc. All these are building up what is called in the USA the "info-culture". They are very dispersed and do not constitute a well defined corporation. But they are very likely going to play a crucial role in the forthcoming civilization.

Design and innovation policy
Misunderstandings which have historically affected design have been apparent in the development of innovation policy, which is quite a recent concept. It appeared in the vocabulary and instruments of policy makers about fifteen years ago. One can distinguish three phases in innovation policy:

(i) During the latter part of the 1960s and the early 1970s, innovation policy was just emerging. It was conceived as an appendix to science or research policy, even when those who tried to promote innovation policy saw it in quite a different way (see for instance the *Charpie Report* issued by the US Department of Commerce in 1967). The promotion of design was in general "forgotten" in the measures that industrialized countries were setting up to stimulate innovation during this period. The bulk of these measures were conceived as aids to development, or as the means for accelerating the application of research results.

(ii) Then there was the decade, approximately 1973 to 1983, during which the concept of innovation policy was gradually institutionalized. The promotion of innovation became gradually distinct from the promotion of R&D. During this period, the promotion of design mainly took the form of "Don't forget it" campaigns. Engineering schools were made aware of the importance of introducing design courses in their curricula. Campaigns were launched to alert industry, and notably small firms, to the role of design in their overall strategy. Steps were taken to justify firms' outlays on design as worthy of receiving support in public grant schemes. Agencies in charge of public procurement were alerted to "value analysis" or "design to cost" methods and related experiments were launched.

(iii) We are now entering a new phase of innovation policy. This policy tends to be seen less as a complement to science policy than as an overall policy to manage technological and social change as a whole, at a time when industrial societies are experiencing deep adjustment of their economies and values. This approach was clearly set up at a recent OECD Workshop (Dubrovnik, September 1983).
Within this overall perspective, innovation policy appears as a

second-level policy. Its role is not so much directly to support the development of new products and processes (and as such to be complementary to other policies). It is rather seen as a means to create a climate favourable to innovation and thus to review, reorient, transform and coordinate existing policies which have been established and are operated in their own right (and not with a view to promoting innovation). This is true for research policy, educational policy, social policy, regional policy... Innovation policy becomes a new tool, a new approach for public management and administration as a whole.

A comparative analysis of innovation systems and related policies in Europe, Japan and the USA (see the OECD's *Statement* at the Six Countries Workshop held at Bonn, 10-11 May 1984, and the OECD *Observer*, November 1984) shows that three key functions should be fulfilled:

(a) The first related to the means of development, i.e. the establishment of a solid base for the accumulation and transmission of knowledge and know-how. The USA exemplifies the role of an advanced research system, while Japan illustrates the key importance of educational efforts through basic schools, media, complementary education, and work-related training (quality circles).

(b) The second function is the establishment of an appropriate institutional and regulatory framework for innovators. This deals with competition laws (anti-trust), competition-cooperation rules (see Kobayashi's paper "Japanese Industrial Policy"), deregulation, and elimination of all sorts of obstacles to innovation and entrepreneurship (bureaucratic, financial, etc.). It also includes the provision of public financial support for innovators.

(c) The third function relates to objectives: it concerns the formulation of focal points or clear targets for structuring technological efforts at national level. These have been provided in the USA, for instance, by big projects in defence and space, and in Japan by "visions" of future technological and socio-economic trends, often resulting in concrete goals such as the fifth generation computer.

It is not easy to figure out what could be the status of design within such new perspectives for innovation policy. Some remarks may, however, be made in relation to the three functions:

(a) Design should be part of the basic "technical culture" of researchers, innovators, entrepreneurs. It should be introduced appropriately at the different levels of the education system, no longer being the domain of specialists, but part of a basic culture to be applied in various innovative practices.

(b) The concern for design can be a useful tool to detect and reveal a number of obstacles to innovation. This applies for instance in public procurement practices or standardization.

(c) As a global targeted approach, taking into account multiple parameters, design as *état d'esprit* and discipline may find a number of applications regarding formulation of nationwide projects. It may also be useful in the elaboration of a national policy regarding external trade (notably the substitution of domestic products for imports) which could efficiently draw upon the techniques of value analysis.

From these remarks it is clear that design should be promoted both as a precise know-how and as a general *état d'esprit*. This requires appropriate measures addressed to various areas of society: schools, business enterprises, administrative bodies and government agencies, etc. It would take too long to discuss here how such measures can be conceived. However, the need for this pervasive approach is urgent as the industrialized societies, particularly the European ones, are coping with deep technological changes which dramatically modify the conditions of competition at world level, and which will lead to the development of a new civilization.

*The opinions expressed in this paper are those of the author
and do not necessarily represent the views of the OECD.*

REGULATION AS A STIMULUS FOR
TECHNOLOGICAL CHANGE

Nicholas A. Ashford

I want to discuss what some policy analysts may think is a tangential issue with regard to the innovation question and government policy, and that is regulation. Certainly in the western industrialized countries regulation is not a tangential issue; and more than that, it may have a broad-based application in the developing countries and newly industrialized countries where demands for quality of life issues are growing as rapidly as the demand for new products and services. Regulation can be used effectively not only to meet a variety of social and economic demands but also to provide a stimulus for dynamic change, as opposed to achieving economic efficiency which is where traditional economic regulation focuses. The remarks I offer include a discussion of both economic regulation and health and safety regulation.

The United States is probably the most regulated advanced industrial system in terms of health, safety and the environment, perhaps closely followed by some countries in western Europe. That regulation has engendered an antagonistic response on the part of industry, antagonistic for a variety of reasons. First, it caused some restructuring. Second, it allegedly diverted resources away from what industry views as proper economic activity for the production of goods and services, and more recently it has been alleged that regulation hampers industrial innovation. Anyone who might have been interested in the regulation field can no longer ignore the challenge to look at the data and provide some insight as to what it is that regulation really does and can do with regard to innovation in the industrial sector.

Now my own interest and the interest of the Center for Policy Alternatives in the effects of regulation on technological innovation came about quite accidentally because I directed, as my first task at CPA in 1974, what was called the 5-Country Study. This was a static terms. And we are talking about *innovation*, which is a dynamic concept.

With regard to the effects of regulation on technological change, we recognize the differences between innovation and the diffusion of technology. In some cases we *want* diffusion of control technology, let us say — widespread diffusion. In other cases we do not want the diffusion of existing technology. We want to use regulation as a technology forcer, as the driving force to produce a new kind of product, such as the substitute for PCBs in electrical transformers.

There is a model of innovation which is useful, and while it certainly does not explain everything, it is a model of innovation without regulation which can serve as a kind of starting point. That is the model proposed by Abernathy and Utterback in which the rate of technological innovation is graphed as a function of time, and it is a peculiar coordinate, time, because this is the kind of graph where the firm could go backwards and forwards. So it is evolutionary time, of a sort. And what was noticed by these researchers (and, with some validity, in a number of different product sectors) is that the rates of product innovation occur differently temporally from the rates of process innovation for a number of industries, whether you are talking about automobiles or chemicals. Product innovation peaks long before process innovation; then both rates eventually diminish as the industry matures.

The question is: when regulation, or any exogenous market stimulus, really destabilizes the system, perhaps sub-optimal to begin with, how does the firm respond? Is it the firm that responds, or is it a non-regulated firm which responds, a new entrant? Do we stimulate the firm to move back up its product line development, or does a new firm enter the picture? You know the story — the significant innovations have very often *not* been pioneered by the displaced firm, and that has nothing to do with regulation. I am talking about radial tyres displacing commercial tyres, electronic calculators replacing mechanical calculators, and of course the vacuum tube producers are not those who successfully pioneered the transistor. So the important question to ask, then, is what is the effect of a regulatory stimulus on either the regulated industry or other respondents. Note that the response can be either a technology to take the firm into compliance, and/or changes in what I call main business innovation — that is, in the process of complying, other kinds of technologies emerge. When asking what is the history of regulation and how can one design regulation to bring about changes that one wants, one asked: Do you want to encourage a regulated firm, or another outside firm, to respond? That may depend upon whether or not you believe the regulated firm has any innovative potential left.

The most important thing about a regulation as to whether or not you get innovation or diffusion of existing technology, or innovation by the regulated firm as opposed to a new entrant, is the stringency of the regulation relevant to the innovative potential of the regulated firm. What do I mean by "innovative potential"? What I mean by innovative potential is very closely related to the pattern of innovation of the regulated industry prior to the regulation. If it is in an innovative state where it is changing product technology as a matter of course, then changing one of the dimensions — safety, health, or whatever — is not particularly difficult and it is that firm that will respond. If, however, you are demanding that product change from a firm that has long since lost its interest in product innovation — you may very well stimulate the new entrant to enter with a new product.

The firm's technology is a major determinant of the technological response. There are better and worse ways of regulating because you can get different technological responses depending on what you do. Intelligent regulation can give you technological responses, some of which are consistent with technological growth and innovation and some of which are not. And, by the way, I think that what this points out is that regulation is not so much the tension between the beneficiaries of regulation and industry, but rather it is a tension within and between competing industrial sectors. It is restructuring that creates the tension, not the tension between health or safety and economic growth.

Industry has claimed that what any response to regulation does is simply to divert capital. It argues that regulation diverts capital from R&D which means it diverts capital away from innovation. We know that R&D is not necessarily linked to innovation. The other half of the story is that diversion of capital is not necessarily diverted from R&D. So in order to really show that regulation hampers innovation, you have got to show that the money is diverted from R&D in the first place, and secondly, that when you do that, it really hampers innovation. Based on empirical evidence, it is a very weak conclusion.

If you look to some of the data which exist in the chemical, pharmaceutical and automobile industries, you find some very interesting things. R&D as a percentage of sales for the chemical industry has been going down steadily since 1960 and you do not see a difference in the slope during the period of high regulatory activity from 1970 to 1980. You see no difference. And it is continuing to go down during Reagan's time. The point is that the chemical industry has realized that it is R&D-rich.

With regard to pharmaceuticals, there is a very interesting set of allegations that efficacy requirements, proving that the drug is effective, have been the reason for the decline of pharmaceutical innovation. If you look at the number of new chemical entities over the past 25 years, there are some interesting things to observe.

First of all, the rate of innovation begins to fall dramatically in 1960, two years *prior* to regulation. Now I do not know any empirical econometric study which will prove cause and effect when the effect precedes the cause by two years, but one could argue that is an artefact of those two years.

But, when you look at the new chemical entities and you weight them by therapeutic importance (and this is important for distinguishing the shift in the kind of innovation that occurs), it turns out that there has been *no* shift in significant therapeutic innovation in the aggregate. However, you must disaggregate pharmaceutical data — because it is very wrong to treat anti-cancer drugs the same as you do anti-infectives, the same as you do anti-hypertensives. These segments are really in a different state of technological evolution, just as fertilizers and polymers are not in the same place. When you do that, what you see, in fact, is a shift in the ratio of prototype to non-prototype drugs. That is, the ratio of the prototypes (the dramatic new changes) to the "me-too" drugs has changed dramatically in favour of the new drugs. One thing that has not happened is a diversion of managerial and technical manpower, which is what industry has claimed has happened as the second adverse consequence of regulation. Instead, we have a redirection. There is a difference between a diversion and a redirection. The diverting and redirecting the technical manpower in the pharmaceutical industry has had dramatically positive results in some firms.

The second thing you see is that there are certain firms whose rate of drug clearance through the FDA process, which is the bogeyman of regulation, has dramatically increased, while others have dramatically decreased. In other words, those companies (and you can identify them) who have sought to match pharmaco-kinetic activity with disease causation models, which is a rational, searching and innovative quest, instead of serendipity, have in fact had a significant payoff in the drug industry.

Next, consider the automobile industry. In the United States during the Carter administration, the conventional wisdom was (and

Detroit said): "We can't do it. We can't meet the safety regulations. We can't meet the environmental emission regulations. We can't meet the fuel economy requirements. There's no way we can do it." And while they were saying they can't do it, Japan and Sweden and Germany were doing it and penetrating the US market. The countries differed in their successes because they had inherited a different tradition of concern for one or more of those areas, but to a large extent they put the US technology in their automobile and reimported it into the United States. *They* innovated; the Americans invented. This caused a subsequent response in Detroit, where for the first time serious reconsideration of both product and process redesign is now occurring.

What implications do these findings have? What consequences do they have for design, the design of regulatory systems for innovation? First of all, you can design regulations to yield the kind of technological response you want. Secondly, there is a need to coordinate innovation and regulatory policies. There is a lot of co-optimizing which can be done. I believe that regulation, though it was not originally invented for that purpose, should be considered as one instrument of technological innovation policy — especially in a cultural environment that is demanding safety, demanding environmental soundness. That can very well be in developing as well as developed countries.

Third, I think you need to build in sufficient uncertainty in the system to cause the kind of destabilization that you want. Too little uncertainty does not give you any innovation at all; that is where you get the conventional wisdom about specification standards not providing an innovative potential. If you are vague and totally unstructured about what your demands are going to be, nobody wants to take a chance to sink capital into a new innovation. Where is the optimal point? I do not know, but you need to build a certain degree of uncertainty — just enough to stimulate the need for technological change. This presumption is consistent with the very good work of Burton Klein, in his book *Dynamic Economics*. He talks about the need always to resist the status quo. Don't let the firms become lazy. In the Utterback/Abernathy language, don't let them get too far along the product evolutionary line — you've got to continue to build destabilization in order to have real economic growth.

Now, what are the lessons for economic regulation? If our aim of economic regulation is to avoid monopoly pricing with unfairness to the consumer, we miss the whole point. The point is to be able to stimulate the kind of dynamic change that we need to have. That requires a different look at the factors having to do with traditional, economic regulation. And whether you are talking about the kind of economic regulation which is reflected in the social concern of labour displacement, or you are talking about straight economic concerns of a capital nature, I think here too you have got to co-optimize. To deal with the issue of labour displacement as an afterthought, at the same time that you are creating a whole set of new incentives for technological growth, is ineffective and potentially costly. You have to open up the engineer's problem space. The way you open up his problem space is to make sure he understands all the dimensions along which he has to design. That includes job redesign and health, safety and environmental concerns.

Bibliography

N.A. Ashford, C. Ayers, R.F. Stone
"Changing the Market for Innovation using Regulation", *Harvard Environmental Law Review*, vol.9, no.2, Summer 1985.

N.A. Ashford, R.F. Stone
Evaluating the Economic Impact of Chemical Regulation: Methodological Issues, prepared for the Netherlands Ministry of Housing, Physical Planning and Environment, February 1985.

N.A. Ashford, G.R. Heaton, U.C. Priest
"Environmental Health and Safety Regulation and Technological Innovation", in *Technological Innovation for a Dynamic Economy*, C.T. Hill and J.M. Utterback, eds, Pergamon Press, NY, 1979, pp.161-221.

N.A. Ashford, G.R. Heaton
Environmental Regulation and Technological Change in the Chemical Industry: Theory and Evidence", in *Federal Regulation and Chemical Innovation*, C.T. Hill, ed., American Chemical Society, 1979, pp.45-66.

N.A. Ashford, G.R. Heaton
"Regulation and Technological Innovation in the Chemical Industry", *Law and Contemporary Problems*, Duke University School of Law, vol.46, no.3, Summer 1983, pp.109-137.

N.A. Ashford, G.R. Heaton
"Regulation and Technical Innovation", *EPA Journal*, September 1979, pp.32-34.

LINKING INNOVATION THEORY
TO INNOVATION POLICY

Tom Casey

> *Technology discloses man's mode of dealing with Nature, the process of production by which he sustains his life, and thereby also lays bare the mode of formation of his social relations, and of the mental conceptions that flow from them.* (Marx, 1954)

The stresses and changes induced in the social and economic fabric of the West over the last ten to fifteen years are more and more recognized as not being of a short-term or cyclical nature but as arising from fundamental change in the economic structure of advanced industrialized countries.

Neither understanding of the changes taking place nor management of the adjustment of structures has been helped by governmental economic policy, which is nearly always derived from neo-classical economic theory.

Whether Monetarist or Keynesian, the large-scale tightening or loosening of the purse-strings, fiddling with monetary or fiscal policy, has been based on theory discredited even in the eyes of its users. Any sign of economic growth creates as many questions as it answers and any attempt to link it to governmental economic policy is a case study in sophistry and politics.

It is in this morass of ethereal opinion and political prejudice that economic policy is often formed. It attempts to shape and modulate financial flows without attempting to understand the driving forces. Identification of variables and questions of causality become essentially a political assumption, and justification of policy a diatribe in prejudice.

Neo-classical theory, its supply and demand variables and its equilibrium concept of growth, is evident in its inability to relate to the tragedy that is Europe today: twelve million unemployed in the EEC, while its wealth-creating base, its industry, becomes more and more suspect under pressure from the US or the Far East.

The relevance (except as limiting factors or boundary conditions of interest rates, money supply or gross demand) to the regeneration of our industrial base (either as new manufacturing or service industries) is in question. Yet this inability to articulate and make operational policy other than in neo-classical macro-economic variables leads to the ubiquitous domination of the Department of Finance within government and the delegitimization of arguments arising from other departments and even from industry itself, whether from the left or right.

Their policy arguments are legitimate only in so far as they can be transmitted into this economic logic and language. Understanding of patterns and mechanisms of causality are lost in aggregation and translation into respectable "economic policy".

The vital need is to examine the mechanisms which drive the changes in the macro-economic, neo-classical variables, not simply to adjust these variables on political whim. The need is to drive such variables as exchange rates, interest rates, factor supply and demand, etc. with mechanisms based on an understanding of industrial and social change at a micro-economic level, where intervention can be based on genuine understanding of patterns of causality and change.

In that development has not and will not follow a constant pattern, with all factors retaining a constant relative importance and set relationship, it seems evident that policy aimed at augmenting and smoothing the development process cannot effectively be based on simple adjustment of macro-economic parameters if they are dependent on sets of micro-economic factors themselves changing in their relationships over time.

The renewed interest over the late 1970s and 1980s in the work of Kondratiev and Schumpeter, as well as the growing recognition of the role of technological change within economic development, signals a move to produce policy dependent on adjustment of variables which drive the macro-economic outer shell rather than vice versa.

Allied to this shift has been an equally pronounced move by industrial economists to start to examine the dynamic nature of industries rather than being content with static descriptions and correlations. Here the driving forces of competition, with continual readjustment pressures building up and being released within industries, has led to far richer concepts of industrial change and development.

The intention here is to examine a number of these developments in the analysis and theory of technological innovation and to relate them to possible orientations and means of generating policy for economic development, particularly regarding innovation policy. It does not claim, however, to provide the link to the surface macroeconomic phenomena, but it may indicate possibilities for further development.

The three main developments in the theory and analysis of technological change examined here are:

1. technology as a competitive factor of firms;
2. the firm as the creator and evaluator of technology;
3. relevant structural definitions of industry and the role of technology in cyclical and structural economic change.

Technology as a competitive factor of firms

This section indicates the interactive and complex nature of the way in which firms, either public or private, deal with technology in ensuring their own survival and development. It starts to lead to the conclusion (reached at the end of the next section) that the firm is, within the non-communist world, the key institution in exploiting and transforming technology and the means of production. This conclusion then leads to a number of policy implications.

The steady reorientation of the analytic direction of industrial economics has brought the work of people such as Porter (1980) and Ohmae (1983) not just to the attention of business studies students but also into economics courses. Analysis searches not so much for correlations between profit and concentration or between firm size and innovativeness, but has moved towards the examination of the strategies and actions of the firms and relating these in an interactive manner to the perceived development of industries. The hubris of much previous quantitative work is being abandoned and the search is for an understanding of the driving forces of competition which are continually reshaping firms and industries.

It should be emphasized that this notion of competition is radically different from the neo-classical economic concept. In the work of Porter, for example, competition is not viewed at all as perfect and it is very much the job of the firm to make it even more imperfect and to its own advantage. Equally the industrial structure has a strong influence in determining the competitive rules of the game. However, while he recognizes that these underlying characteristics of an industry are rooted in its economics and technology, the latter hardly enters into his methods of analysis. Thus, competition employing cost leadership, differentiation or focus takes place within a specific sector's structure itself, generated by the interaction of the firms both with each other and with the underlying factors.

Porter presents mechanisms both for the development of the firm and for the reshaping of the industry. They emphasize the central role of the firm and elaborates methods of generating individual firm strategy.

Competitive technology strategy in firms is not yet as well developed as general competitive strategy or other functional strategies such as marketing.

Freeman (1982), in his chapter "Innovation and the strategy of the firm", has indicated various possible forms of strategy and how the firm uses and may try to change its technology.

Ford (1983) treats technology very much as another competitive strategy. Here technology is bought and sold depending on the firm's other abilities and resources, the market size, the nature of the competition and other restrictions. There is a healthy demystification of technology in Ford's work, even though it does not have the sophistication of Freeman's. The emphasis is on generating a strategy which will produce a return and will strengthen the firm. The approaches of both Freeman and Ford imply that firm innovation policy be closely allied to other functions of the firm.

The active exploitation of technology as an important factor in the survival and development of the firm is now being integrated into the previous work. Here the use of technology, its change, its adaptation and its exploitation, has become an important element in the perception of firms. The treadmill of constant technical innovation is the modern curse of Sisyphus on firms. Competitive technological strategy has become a key area in firms' planning as has its interaction with other functions of the firm (marketing, finance, training, etc.).

Innovation in new technology, whether product or process, based on in-house or contract R&D, licensing joint venture, acquisition, or simple purchase, forms part of the firm's overall strategy for dealing with a changing economic environment and competitors. The technology and changes attempted in the technology must be appropriate in the position from which the firm is competing. It is evident that purchasing high-productivity equipment may not be appropriate when the market is rapidly changing and requires a high degree of flexibility etc. Equally, technical innovation strategy needs linking to other strategies such as management or training.

In regard to government industrial policy, this type of work is leading to a greater initial examination of the competitive requirements of different firms in specific industries, rather than the more directive policy which assumes a longer-term knowledge of how the industry should develop and thus what sort of firm should be encouraged.

It is evident from the start that policy must be on a sectoral basis. The competitive situations and needs of firms are so different from industry to industry that the broad-brush approaches cannot possibly be satisfactory. Then analysis of the competitive structure of the industry must be the critical starting point of its technical innovation policy. This analysis must seek out not only what tend to be called the "key success factors" but also indicate their dynamic nature and the interdependencies which may exist between them.

The analysis of technology on its own is not sufficient. Policy directed at the narrowly defined technology function within a firm is also not sufficient. Firstly, the policy should be coherent and consistent with the appropriate technology strategy of the firm for improved competitiveness. Secondly, the relationships to the other functions of the firm need to be recognized. That is to say, such factors as government procurement policy, education and training policy, tax policy, R&D policy etc. need to be considered in relation to technical change within the firm. For example, changing technology in a firm may be very closely linked to increasing capital intensity and cash flow/working capital requirements and to radically different human resources requirements which cannot be treated separately.

The firm as a creator and evaluator of technology

The importance of the role of the firm in exploiting technology for growth has become evident in the last section. This section goes further to indicate the firm as the creator and evaluator of technology and thus to highlight its key role in economic development.

The last section has seen technology being exploited by the firm as a competitive strategy, as a means of survival and development. It can now be asked what the effect of this process is on the technology itself and on the industry as a whole.

There are, of course, certain physical boundaries within which the technology is formed. However, the areas in which technological change is developed or not depend to some extent on the competitive relationships between firms (not only in terms of collusion or oligopoly effects but also within a positive Schumpeterian framework) and the relationships of firms to their industrial structure and general socio-economic environment. This is one of the thrusts of economic growth which firms give to the

indicate underlying relationships not apparent as significant elements in simple input/output analysis. Here motivation may rest on factors outlined by industrial economists from Bain to Porter; guarantees of markets or supplies, rationalization, economies of scale, etc.

Yet other articulations between groups of firms may exist in the developing technological relationships and dependencies between firms. Moving from batch processes to continuous process has often defined axes along which firms have developed, as has the continuing integration of the functions of firms often apparently distant. For example, integrated circuit manufacturers have "integrated" the *raison d'être* of other firms' functions onto their chips with a consequent structural reshaping of the industry. Software firms threaten to carry out similar exploits in the service industries with sophisticated applications software and expert systems.

These three types of articulation of exchange, of ownership and of technology may at the same time be defining different complexes within the economy and different possibilities for structural change as the relationships between groups of firms.

It was remarked in the previous section that firms were involved in the commercial exploitation of technological change within a set of parameters such as technological opportunity and appropriability. Nelson's natural trajectories can now be related to the technical articulations and can be seen as paths of development along which exist technological opportunity. These trajectories may be generated externally to the industrial grouping of firms and may be adopted by them (e.g. microelectronic components, computer control etc.) or they may be generated internally (specific improvements or integration of the technologies).

The economic exhaustion of these paths of technological opportunity may well cause the grouping of firms to fall into the "commodity" league, with the key competitive factors moving towards factor prices. Renewed technological opportunity from external sources may engender rapid switching as regards suppliers (metals to plastics, mechanical to electronic, etc.) with reorientation of the articulations and consequent structural change. This search for paths of technological opportunity in which the firm can create a form of technology which will be to its competitive advantage is then a key element in industrial/economic structural change and development.

In terms of government policy generally, strategically articulated groups of firms/sectors (where the maintenance of the linkage may be of competitive importance) might be identified and supported to avoid a "domino" effect on the competitiveness of all groups. This interdependence may lie along any of the three types of articulation indicated.

Another possibility is that the externally provided competitive technological opportunity may be of such importance to so many groupings of firms that it is essential to have a capability therein.

References

Ford, D. (1983)
> "The management and marketing of technology", paper given at Strategic Management Society Annual Conference, Paris, October 27-29.

Freeman, C. et al. (1982)
> *Unemployment and Technical Change*, Frances Pinter, London.

Marx, K. (1954)
> *Capital*, vol.1, Lawrence & Wishart, p.352 fn.

Nelson, R.R. (1980)
> "Production sets, technological knowledge and R&D", *American Economics Journal*, pp.62-7.

Nelson, R.R. and S.G. Winter (1977)
> "In search of a useful theory of innovation", *Research Policy*, 6, pp.36-76.

Ohmae, K. (1983)
> *The Mind of the Strategist*, Penguin.

Porter, M.E. (1980)
> *Competitive Strategy: Techniques for Analyzing Industries and Competitors*, The Free Press, New York.

PUBLIC INNOVATION POLICY: TO HAVE OR TO HAVE NOT?

Roy Rothwell

For many years governments in the advanced market economies have had policies designed to stimulate scientific advance and technological change. These have included funding basic research in universities and government laboratories, technical education and the establishment of a system of collective industrial research. Policies enabling individuals and industrial companies to appropriate the benefits of their inventive and technological development activities have also been promulgated, specifically national systems of patenting. Alongside these, in a number of countries governments have established institutions to assist in the transfer of publicly generated patentable technological change to commercial use, examples being the NRDC in Britain, the JRDC in Japan and ANVAR in France.[1]

Similarly, governments have for many years been involved in formulating and implementing industrial policies. These have included policies towards industrial restructuring, tariff policies, capital grants and corporation tax regimes. While in some countries governments have played a direct, indicative role in influencing the operation and structure of industry, e.g. in France and Japan, others have adopted more of a hands-off approach towards industry, e.g. in the United States.

More recently, during the late 1970s, governments began to formulate policies designed to stimulate increased rates of industrial technological innovation; in other words they began to adopt explicit *innovation policies*. If we define innovation as "the technical, financial, managerial, design, production and marketing steps involved in the commercial introduction of a new (or improved) manufacturing process or equipment", i.e. a process which involves the whole gamut of activities from invention through to commercialization, then we can see that innovation policy is essentially an integrative concept. In other words, innovation policy represents the combination, in a coordinated manner, of science and technology policy and industrial policy.

When we look at innovation policies in a number of countries we can see several important differences in the nature of the policy process, in the policies adopted and in the types of policy tools employed. Taking first the policy process, in the United States it is highly "political", decentralized and involves a great deal of inter-group negotiation and bargaining; partisanship (strong lobbies) and high visibility appear to be important to policy stability.[2] At the opposite end of the spectrum, policy making in Japan is based on public and private consensus with strong central direction; policy stability tends to be high since, given the political situation in Japan, innovation policy making effectively is largely apolitical. In Europe, the policy process varies between nations, but lies between the two extremes of the United States and Japan.

Not surprisingly, the national differences in approach to innovation policy reflect the roles that governments play in the economy and in industrial development generally, and in this respect two quite distinct kinds of state intervention can be discerned:[3]

(a) In some countries state intervention in industry is seen as a major part of a process of indicative planning. This is the case in countries such as, for example, Japan, France and Italy, where industrial policy is used as an important instrument for economic policy and where the objectives of that policy are formulated within a framework of economic and social development plans, which are indicative for the private sector. Innovative policy is then formulated through consultative and coordinative procedures and institutions within government and between government and industry.

(b) In other countries industrial policy is seen as part of general economic policy, aiming to create a favourable climate for industrial development. Although these countries, like the Netherlands, Denmark and the German Federal Republic, use industrial innovation policy instruments or even sectoral policies, these policies are not formulated within the framework of a national plan, nor are they used as selective policies in an intensive or systematic way.

The above distinction of the two main ways of formulating industrial policy should not be seen as a model for describing two totally different worlds. Often, differences are not as great as they would appear. Of course, preferences regarding the various economic goals of political parties in power play an important role.

Whereas full employment and equalization of income distribution can be considered as traditional priority issues of socialist-labour parties, conservative parties traditionally place more emphasis on price stability and balance of payment equilibrium. These different aims, clearly, will influence the kinds of industrial and/or innovation policies adopted.

The kind of policy approach adopted by a government will considerably influence the degree to which public funds are directed into certain industries or areas of technology, i.e. the degree of technological selectivity.[4] It is perhaps in Japan that the level of selectivity is greatest in both the public and private sectors, which collaborate closely in major national projects in such areas as amorphous semiconductors and fifth generation computers. Projects are selected following a consensus-obtaining, opinion-sampling exercise, in which MITI plays an important coodinating role. Resources are then focused on to the selected technologies in both national laboratories and the R&D laboratories of major companies, and the projects are supported through both public and private funds (see Table 1). Other countries (and the European Community) are now copying this approach.

There is also a high level of selectivity exercised by the French government, which presently controls over 50% of national industrial R&D and research workers (see Table 2). The recently enacted Programme Law, designed to provide encouragement and incentives to industry to increase its R&D commitments, defines a number of "mobilization programmes" for R&D, which apply to both the public and private sectors. The government has also selected five major industrial areas to be preferentially supported from public funds: steel, chemicals, electronics, health and materials. A significant feature of the French system is the concept of *filière*, which applies to the chain of production from, e.g., raw materials to the commercialization of final products.

In the German Federal Republic, where the bulk of industrial R&D is carried out in industry with its own funds, it is industry that largely established its own R&D priorities. The BMFT generally acknowledges that it should only become involved in projects with certain special features, such as especially high scientific and technological risk, very large financial requirement, long development horizon, lack of short-term market acceptance but high public benefits. The choice of rather broad areas for support is made by BMFT, but in consultation with industry, professional associations through an elaborate advisory system, and, of course, parliament.

Table 1   Areas of interest to Japanese industry

New Products	Energy industries	Advanced, high-technology industries
Optical fibres	Coal liquefaction	Ultra-high-speed computers
Ceramics	Coal gasification	Space developments
Amorphous materials	Nuclear power	Ocean developments
High-efficiency resin	Solar energy	Aircraft
	Deep geothermal generation	

(*Source*: Japanese Ministry of International Trade and Industry)

Table 2   Strategic high-technology priorities in France

Strategic Industry	Objectives	Overall actions planned
Electronic office equipment	To achieve 20–25% world market share, and avert an anticipated $2bn trade deficit in 1985	In strategic sectors, the government will negotiate development contracts with individual companies, setting specific goals for sales, exports and jobs. Firms that make such commitments will receive tax incentives, subsidised loans, and other official aids.
Consumer electronics	To create a world-scale group including TV-set and tube makers that will each rank among the top three globally. To eliminate the $750 million trade deficit in such products.	
Energy-saving equipment	To ensure that government grants to companies and households to install such equipment are spent primarily on French products.	
Undersea activities	To recapture second place in the world after the USA.	
Bioindustry	Objectives not yet defined.	
Industrial robots	Objectives not yet defined.	

These six industries together are expected to add $10bn in sales and to double their workforce to 135 000 by 1985.

(*Source*: *Business Week*, 30 June, 1980, p140.)

In the United States, governments traditionally have had an aversion towards attempting to "pick winners", relying on market dynamics to dictate efficient patterns of R&D allocation. In reality, however, defence and space R&D support and procurement activity inevitably have biased a large proportion of the national R&D effort along certain specific paths and have effectively stimulated the growth of new technology-based industries. In a significant departure from established practice, the Carter administration suggested the provision of public support for selected generic technologies. This was abandoned by the Reagan administration, who instead have attempted to facilitate the establishment of generic cooperative research activities by industry itself, the first of these being the cooperative effort in microcircuitry by the Semiconductor Industry Association in California. In practice, in the USA (as well as in Europe) market dynamics have forced the administration to devote considerable resources to "bailing out" problem industries (e.g. the automobile and shoe industries) rather than to supporting high technology industries (apart from working through aerospace and defence budgets).

In the UK, in the past, there have been some attempts to support developments in specific high technology industries (e.g. computers and nuclear energy), but since the mid-1970s the British government has adopted a "market forces" policy that is largely non-selective. Just as in the United States, however, the high level of public R&D support for defence-related projects has invariably focused resources on specific areas of technology. In addition, and again following the pattern in the USA (and other countries), a large portion of public funds has gone towards supporting ailing industries, most notably automobile and steel production. More recently, the British government has become overtly selective and, following the Alvey Report on a programme for advanced information technology, has released plans for the selective support of major collaborative projects in this area. This is likely to be followed by a second programme in the area of advanced materials technology.

A number of the above differences can be illustrated through consideration of Table 3, which is an analysis of innovation policy recommendations by type of tool in six advanced market economies. The data are taken from national innovation policy statements published in the late 1970s.[5]

In Britain the greatest emphasis is on financial and taxation measures, which appears to reflect a preoccupation with obtaining a healthy environment for industry to operate in. In the case of the

Table 3   Classification of government policy tools

Policy tool	Examples
Public enterprise	Innovation by publicly owned industries, setting up of new industries, pioneering use of new techniques by public corporations, participation in private enterprise.
Scientific and technical	Research laboratories, support for research associations, learned societies, professional associations, research grants.
Education	General education, universities, technical education, apprenticeship schemes, continuing and further education, retraining.
Information	Information networks and centres, libraries, advisory and consultancy services, data bases, liaison services.
Financial	Grants, loans, subsidies, financial sharing arrangements, provision of equipment, buildings or services, loan guarantees, export credits.
Taxation	Company, personal, indirect and payroll taxation, tax allowances.
Legal and regulatory	Patents, environments and health regulations, inspectorates, monopoly regulations.
Political	Planning, regional policies, honours or awards for innovation, encouragement of mergers or joint consortia, public consultation.
Procurement	Central or local government purchases and contracts, public corporations, R&D contracts, prototype purchases.
Public services	Purchases, maintenance, supervision and innovation in health service, public building, construction, transport, telecommunications.
Commercial	Trade agreements, tarrifs, currency regulations.
Overseas agent	Defence sales organisations.

(*Source*: Rothwell and Zegveld, 1981)

United States, 50% of the measures are in the category "legal and regulatory", which reflects a deep concern about the economy being over-regulated.[6] These two contrast with the other four countries which prefer to deal more directly with the inputs to the process of innovation. Thus in the United States and in the United Kingdom policy emphasis seems to be on creating a climate conducive to firm-based innovatory endeavours, government leaving the choice of technology in the hands of individual managers, market forces dictating patterns of resource allocation.[7] Other countries, most notably Japan and France, have clear-cut, long-term strategies towards the development and exploitation of specific high technology product groups and new technologies (Tables 1 and 2). Such policies have a strong reindustrialization flavour, being aimed essentially at the structural transformation of industry to higher value added, more knowledge-intensive sectors and product groups.[8]

Are national innovation policies necessary?
According to Roessner,[9] it is highly unlikely, in the absence of a deep national crisis, that a comprehensive and coherent national innovation policy could be undertaken in the United States. This is because of the nature of the American political system with its deep-rooted suspicion of centralized decision making and indicative planning. In underlining his point, Roessner quotes Dahl:[10]

A central guiding thread of American constitutional development has been the evolution of a political system in which all the active and legitimate groups in the population can make themselves heard at some critical stage in the process of decision.

Despite the lack of a formal, comprehensive innovation (or industrial) policy, the US manufacturing sector nevertheless underwent a remarkable structural transformation during the 1950s and 1960s. This was the result, at least partially, of radical innovations resulting in the creation of major new product groups and the emergence of a new industrial sector (e.g. semiconductors, computers, pharmaceuticals, composite materials, computer aided design and satellite communications). In most cases these radical new industries emerged sooner and grew more rapidly in the United States than elsewhere.[11] Since the transformation occurred in the absence of formal innovation/industrial policies, it is worth while posing the question "are such policies important to enhancing national rates of innovation and international competitiveness, or can we get along as well without them?"

An interesting feature of the emergence of several of these new industries in the United States was the crucial role played by both military and space agencies. In the case of the US semiconductor industry, for example, the Department of Defence played its part both on the supply side (R&D funding) and on the demand side (procurement). This has been summarized by Dosi as follows:[12]

On the supply side the DoD influences involved:

a) Stimulation towards precisely defined technological directions and areas in which to allocate R&D efforts.

b) The incentive toward and the direct financial support of the exploration of different possible alternative paths of technical change. Especially when precise "trajectories" are not yet well defined there is what one could define as the "burden of the first comer" (together, of course, with the rewards for the first comer). This burden, made of the attempts of trying and testing a number of possible areas of advance (which is much greater than that required from imitating industries) has been — partly — undertaken by public institutions (military and space agencies).

c) The speeding up of technical progress at what we would define as the "maximum rate" compatible with the existing knowledge, technology and experience.

d) The subsidy of expansion of productive capacity to certain target levels considered necessary for national defence requirements.

e) A push towards standardization of production.

f) The lowering of entry barriers for new firms which could find a market with low "cost of access".

On the demand side public policies resulted in:

g) A guarantee of a future market for any innovation corresponding to the required technological features.

h) The expansion of demand, with associated powerful learning effects upon productivity and unit costs.

i) (Possibly) a subsidy element involved in public contracts which helped to cover fixed costs (such as R&D) that would otherwise have fallen upon civilian sales.

The policies of procurement, R&D financing and explicit indication of the required direction of technological advance, together operated as a widespread and finalized allocative mechanism, of both productive and research efforts.

Similarly, the US commercial aircraft industry[13]

has been a major beneficiary of government policies that (prior to 1978) simultaneously affected the demand for and the supply of technological innovations. In many ways, the pre-1978 policies toward the US aircraft industry resemble those observed in such Japanese industries as computers and semiconductors. This unique position of this industry reflected the close technological and financial links between the military and commercial aircraft industries.

In the case of satellite communications, Teubal and Steinmuller[14] have described the crucial role of NASA in the emergence of the US civilian industry. In the first case NASA underwrote the cost of development of first passive and then active communication satellites. Subsequently, as part of the ATS programme, NASA developed a series of radically new ideas, thus effectively underwriting risks across a very broad range of innovations. At the same time, ATS succeeded in making a number of notable technological contributions to satellite communications.

Teubal and Steinmuller have estimated that without NASA involvement the commercial satellite communications industry in the US would have emerged in 1970 at the earliest, rather than in 1965, at which date satellite communications were more efficient than alternative cable systems. By 1970, however, improvements in cable technology were such that cable systems would have been more efficient than the newly emerging (effectively 1965 vintage) satellite systems which would have been introduced, in the absence of NASA participation, by the later emerging industry. Given this factor and given the considerable technical and financial risks involved in the development of active geosynchronous satellite systems, the US civilian satellite communications industry — and in turn the industries in other countries — might have experienced further considerable delays in development.

NASA's contribution was divided by Teubal and Steinmuller into three stages: invention, demonstration of feasibility, and commercial development. The invention stage included a series of critical experiments, which determined that no physical barriers existed to orbital transmission and reception. The demonstration of feasibility stage included the testing of a wide variety of system designs which had crucial knowledge (and skill) creating implications for emerging supplier companies. Commercial development came about following the establishment by NASA of two quasi-private companies, COMSAT and INTELSAT, when commercial users began to undertake the contracting of commercial satellite communication systems and ground installations. Regulations favouring an "open skies" policy contributed to the rapid diffusion into use of each system. The crucial role of NASA as the major coordinating mechanism between innovation stages and between supply and demand is obvious.

Finally, based on detailed studies of the development of six industries, Nelson[15] has suggested that both public procurement and government regulations have played an important role in influencing industrial innovation and industrial change in the United States during the past thirty or so years.

We are thus faced with an apparent dilemma. On the one hand Roessner provides evidence and argues persuasively in favour of the hypothesis that (outside periods of national crisis) formal and comprehensive national innovation policies cannot exist in the United States. On the other hand, the results of empirical research suggest that public policies and institutions have played a crucial role in the US in stimulating radical innovations and the emergence of new sectors of industry. The resolution of the dilemma lies in the fact that the successful policies were neither "formal" nor "comprehensive".

In the first case, the policies generally were not formally directed towards the stimulation of civilian innovation. Rather, they were directed towards the development of advanced weapons systems and other devices for public-sector use, the civilian application and knowledge accumulation coming second and being a spin-off phenomenon. In the second case, the policies were not part of a comprehensive national plan, centrally devised, to stimulate higher rates of innovation, but were instituted by particular public-sector institutions to meet specific (usually military) needs. This at least would be an explanation acceptable to public policymakers in the United States.

But it is pertinent to ask the question: "Has the US government been involved in covert — informal rather than formal — policies towards industrial technological change?" According to Rothwell and Zegveld,[16] the answer to this question is a tentative "yes". In addition to creating a considerable source of civilian R&D funding and acting as an innovation-stimulating, risk-accepting market, the US (indirect) involvement in stimulating civilian technology has enabled the various administrations to avoid the dogmatic issue of direct (formal) governmental involvement with industry via the formulation, centrally, of comprehensive national innovation/ industrial policies.

While in the past the DoD may have refuted the above contention, more recently DoD spokesmen have begun to refer to civilian applications per se and the importance of maintaining a competitive advantage *vis-à-vis* other countries, notably Japan. For example, in 1983 the Undersecretary of State for Research and Engineering, Dr Richard De Louer, is reported to have said: "The 'nth generation development program is the US' answer to Japan's government-supported fifth-generation computer program." The US programme to which Dr De Louer referred (a DoD programme) is formally entitled "Strategic Computing and Survivability".

For the fiscal year 1983 it has been estimated that total R&D expenditure in the United States was about $84 billion, of which approximately 50% was funded by government. About 70% of government R&D was expended under the heading "defence and space". Estimates for the amounts committed to the DoD budget for Research, Development, Test and Evaluation (RDT&E) in the fiscal years 1983 and 1984 are $27.8 billion and $28.6 billion respectively.[17] It would be surprising, given such a high level of expenditure, if the DoD's RDT&E activities did not influence significantly the rate and direction of industrial-technological change in certain areas (e.g. computing, semiconductors, communications, optical systems).

There are, in fact, a number of paths within DoD leading to new civilian technology and an improved industrial base, namely:

- The support of research and development leading to civilian as well as military applications.
- The support of research and development leading only to military applications but providing experience with advanced equipment of interest from a civilian standpoint.
- The allowance of funds for Independent Research and Development (IR&D) in areas of interest to DoD as a cost factor in determining overhead rates for contracts.
- The purchase of large quantities of equipment incorporating advanced technology and requiring sophisticated production techniques.
- The support and the encouragement of university research.
- The implementation of an "action plan for the improvement of industrial responsiveness", which includes as its objectives a sufficient supply of skilled labour to meet the needs of industry and the improvement of industrial productivity.

The overall responsibility of DoD research and development lies with the Undersecretary for Research and Engineering. It is interesting to note that within his organization there is an Assistant Deputy Undersecretary for Production. The latter's responsibilities include productivity and resources. In addition to the armed services, various DoD agencies support research and development. The most important group from the standpoint of new technology is the Defense Advanced Research Projects Agency (ARPA). ARPA serves as the cell for initiating new technologies. After a certain stage of development is reached, the responsibility for programmes is usually transferred to one or more of the armed services. The total budget for ARPA for the fiscal year 1983 amounted to $691 million.

Below are given the results of an effort to identify DoD programmes having potential civilian applications and designed to strengthen the scientific/technical and industrial base (the data are taken from *Reindustrialization and Technology*).[18] The investigation was complicated by the fact that numerous programmes are classified and adequate descriptions of some of the unclassified work is often lacking. The following rough estimates of funds in the fiscal year 1983 for programmes likely to produce new civilian technology are given below:

```
Microelectronics and computers            300,000,000
Data and voice communicating
(incl. navigation)                        300,000,000
Aircraft and hovercraft                   600,000,000
Materials                                 100,000,000
Miscellaneous                             400,000,000
Research, primarily at universities       600,000,000
TOTAL                                   $2,300,000,000
```

Because of the difficulties alluded to earlier in the paragraph, the figures are uncertain and are more likely to be too low than too high.

Of even greater interest is the DoD funding of Independent Research and Development (IR&D) by its industrial contractors, which is estimated at $1.3 billion in the fiscal year 1983. Although some of the IR&D money is already included in the figure of $2,300,000,000 given in the table above, much of it is not. When all these factors are considered, a likely lower limit of the funding of programs with potential civilian applications is believed to be $3 billion in the fiscal year 1983. The actual figure may be considerably higher. When this figure and the programmes to modernize university laboratories and to improve the industrial base are considered (see below), the magnitude of civilian-related activities by DoD is indeed impressive.

Support for universities
All military services and some of the defence agencies, primarily the Defense Advanced Research Projects Agency, have programmes entitled "Defense-Research Sciences". With a few minor exceptions, the activities discussed in the previous section are not included in these programmes. They therefore represent a large source of additional research funding, primarily for universities. The appropriations for them are shown below:

```
                    FY1981          FY1982          FY1983
Air Force           125.2           136.5            155.0
Army                125.0           157.0            182.0
Navy                225.0           255.0            283.0
Defence Agencies    100.7           100.0(est)       101.0
   Totals           575.9           648.5            721.0
```

Because of the basic nature of the work which is supported, most of it probably has long-term potential civilian applications. Eighty per cent is probably a reasonable figure for the share of this type of activity. It is thus estimated that something of the order of $550,000,000 to $600,000,000 is presently supporting research, primarily at the universities, of interest from a civilian standpoint.

In addition, the Lincoln Laboratory (Electronic Systems) is operated for the Air Force by the Massachusetts Institute of Technology. Although much of this laboratory's work is only of interest for weapons systems, some of it certainly has civilian potential. Appropriations for advance technology development at the Lincoln Laboratory amount to $23,079,000 in the fiscal year 1983.

Other activities to strengthen the universities are:

- Beginning with the fiscal year 1983, the military services requested $30,000,000 per year over a five-year period to assist in the modernization of university equipment.
- Industry is encouraged to cooperate with universities by considering this factor in the evaluation of independent research and development (IR&D) when awarding contracts to industrial firms. The IR&D programme is discussed in the following section.
- A fellowship programme for scientific and engineering students has been instituted.
- In the case of the Reserve Officer Training (ROTC) programmes of the Air Force and the Navy, 80% and 70% respectively of the ROTC scholarship holders are required to

enrol in scientific or engineering curricula.

- A cooperative programme provides opportunities for students to work at Department of Defense Laboratories on a part-time basis.
- The Intergovernmental Personnel Act, which is utilized by the Department of Defense as well as other government departments, permits university-government exchanges for periods up to two years.

Improving the industrial base

As in the case of universities, the funding discussed earlier, entitled "Research and Development Programs with potential civilian applications", serves to improve the position of American industry. An additional incentive provided by the Department of Defense is the support of independent research and development (IR&D) on the part of its contractors.

Some or all of the costs of IR&D, with the exception of general and administrative expenses, are allowed in determining overhead rates under certain conditions, namely:

(i) The IR&D must bear some relationship to a military function or operation.

(ii) If the company receives payments of $4,000,000 or less from the Department of Defense in a given year, the negotiation of an advance agreement for IR&D is not necessary. In the case of individual profit centres which contract directly, the limit is $500,000 for each. The costs are then allowed according to a formula based upon the historical average of all IR&D performed by the company or the profit centre to its total sales.

(iii) For companies or profit centres receiving payments exceeding the limits just mentioned, an advance agreement must be negotiated with an Armed Services Technical Evaluation Committee. This committee evaluates the IR&D in terms of its value and establishes the level of support.

It is to be assumed that most companies plan their DoD-supported IR&D projects in such a way that the results are of dual value. Therefore a large fraction of the funding is probably of interest from a civilian as well as a military standpoint. In testimony before a Congressional Committee on the FY-1981 budget, the Undersecretary of Defense for Research and Engineering estimated that annual corporate costs of approximately $2 billion for IR&D were offset by DoD reimbursements of $700,000,000. If these reimbursements have increased in the same ratio as the DoD budget for RDT&E, they should amount to approximately $1.3 billion in the fiscal year 1983. Congress has become concerned about the IR&D costs (also about bid-and-proposal costs, which may be included in overhead) and has directed that they be shown as a line beginning with the FY-1985 budget. For FY-1984 DoD must submit proposed ceilings for the costs, which are to be approved by Congress.

Finally, the Department of Defense is working to increase the productivity and the product quality of its industrial contractors under its Action Plan for the Improvement of Industrial Responsiveness. Programmes of particular importance for this purpose are:

(a) The Manufacturing Technology Program is designed to improve industrial productivity and responsiveness. The activities under this programme, which are funded primarily as part of procurement contracts, are expected to produce "factory floor applications of productivity enhancing technology".

(b) The Technology Modernization Program is handled on a joint-venture basis with industry. Based upon a negotiated agreement which is linked to procurement contracts, DoD invests "in enabling manufacturing technologies and industry investments in capitalization for modernization".

The funding for the programmes could not be determined. When account is taken of the programme description and the possibilities for depreciation changes in contract overhead, the amounts are probably considerable.

The above data have been presented in some detail to illustrate that while the US government might not influence civilian technological change directly via formal national innovation policies, it has exercised a considerable indirect influence on both the rate and direction of civilian industrial technological change via its defense RDT&E expenditures, as well (as suggested earlier) as through defence and space procurement activities. Further, defence RDT&E expenditures increasingly are being utilized overtly to influence civilian technical change, and top-level military spokesmen now openly claim to be consciously fulfilling a MITI-like role in funding and coordinating a significant proportion of US R&D activity.

It may be, of course, that funding civilian R&D via defence agencies is an inefficient means to attain civilian market-oriented ends, and that there are large opportunity costs associated with the considerable level of military R&D. Thus while in the 1980s total US government R&D funding increased ($+2.7\%$ in 1980/81; $+0.6\%$ in 1981/82; $+5.2\%$ in 1982/83), the civilian allocation fell

(-6.2% in 1980/81; -10% in 1981/82; -5.3% in 1982/83). "Overall growth in the eighties was due exclusively to a large increase in defense funding which outweighed the drop in civilian allocation."[19]

A recent investigation of some of the influences of defence R&D spending and procurement activities on the civilian sector in the US highlighted a number of important problem areas:[20]

(i) There is evidence to suggest that the military industrial complex is depriving the civilian market-oriented sector of many of the country's top brains in such crucial areas as computing and electronics.

(ii) Large amounts of human capital are being "locked in" to rather esoteric projects which do not have significant apparent civilian spin-off potential.

(iii) Defence procurement is "feather-bedding" some suppliers (e.g. in the clothing area) due to inadequate quality control procedures and lack of accountability.

In short, it is being suggested that military R&D spending is seriousy distorting the broad thrust of American technological development along paths dictated by military requirements and away from the current and future needs of world markets for civilian goods. This contrasts sharply with trends in Japan. On the other hand there is the possibility that military R&D might be establishing some of the technological trajectories of the future. In addition, there will surely be considerable civilian spin-offs and an increase in the accumulation of national expertise in areas of advanced technology and manufacturing know-how, but whether these benefits will outweigh the costs, or vice versa, remains to be seen.

Finally, it is worth emphasizing that while the US government does not have formal, comprehensive innovation policies and strategies, this does not mean that it is not at all involved directly in stimulating industrial innovation. As Colton et al.[21] have pointed out, the publicly funded National Science Foundation (NSF) has a series of programmes designed explicitly to stimulate industrial innovations which in FY-1984 had a total allocation of $16.6 million. These are the Industry/University Cooperative Research Projects programme, the University/Industry Cooperative Research Centers programme, the Small Business Innovation Research programme, and the Productivity Improvement Research programme. In general, direct innovation-oriented programmes in the US have been piccemeal and have involved relatively minor funds.

At the opposite end of the spectrum to the United States is Japan and, as Kobayashi has described in his paper in this book, the government, via MITI, is involved in formulating explicit national strategies towards innovation and industrial restructuring. Thus, in the case of Japan, the remarkable industrial transformation of the past thirty years was the result of deliberate public policy. Following the second world war, the decision was made in Japan deliberately to restructure its industry better to match the evolving market and technological requirements of the third quarter of the twentieth century. This process has been described by Allen:[22]

A corollary of Japan's determination to lift her productive capacity to a higher plane of technical and commercial competence was her awareness of the need to adapt her industrial composition to changes in markets and techniques. Structural adaptability was recognized as a condition of the continuous growth in GNP. Moreover, Japan did not simply respond to exogenous forces; she was remarkably successful in anticipating change. In the early 1950s her chief export industries still consisted of labour-intensive trades, where her low wages made her an effective competitor with her western rivals, and her superior management and organization kept at bay challenges from the developing countries. But, as her policy showed, she fully realized these advantages were transient, and she soon began to set up a new capacity in several large scale, capital intensive trades, notably steel, shipbuilding and chemical fertilizers. By the early 1960s, when these trades were well established, she turned her attention to a number of engineering industries, especially radio, television sets and motor vehicles, as well as to petrochemical products. By the end of the 1960s her motor car, electronics and watch and clock industries ranked with the world's leaders, while her eminence in steel, motorcycles and shipbuilding remained unassailed. The check to her industrial growth after 1973 was followed by a recovery (in 1978) which was associated especially with the further development of her motor, machinery, instrument and electronic manufactures.

The structural transformation of Japanese industry can be depicted in Figure 1, which is a chart used by the Japanese Economic Planning Agency to explain Japan's economic development. It well illustrates Japanese movements towards higher value added, more knowledge intensive sectors. Moreover, unlike the United States, where reindustrialization involved the initiation of *new* techno/economic combinations, in Japan reindustrialization was based on the acquisition and subsequent improvement of *existing* technological know-how (mainly from the USA); in other words, Japanese reindustrialization was based on a process of rapid technological "catching up".

Discussion

Returning to the question posed in the title of the article, "Innovation Policy: to Have or to Have Not?", it seems that for a variety of reasons there is no simple answer. If the question were reformulated more precisely to "Formal, Comprehensive Innovation Policy: to Have or to Have Not?", then in the light of the evidence from the United States it would appear that such a policy is not essential to achieving high rates of industrial innovation and the creation of new technology-based industrial sectors. Indeed, given the socio/political situation in the United States, it would seem that the adoption of such policies, in the absence of a deep-rooted national crisis, would be impossible.

On the basis of the evidence we have presented above, however, it would appear that informal (covert rather than overt) policies can and do exist in the United States. In general, the aims of these indirect policies are satisfied largely through the extremely high levels of public defence and space R&D expenditure and defence and space procurement activities. Certainly the results of a great deal of defence and space R&D activity in the USA have "leaked out" and found civilian applications.[23] While there might be inherent inefficiencies in the use of indirect policies of this nature it may be that politically they are the only policies that are acceptable in the United States context.

In Japan, in contrast, formal, direct and comprehensive innovation and industrial policies have been crucial to the dramatic shift of Japanese industry to higher value added, more technology-intensive product groups and to the greatly enhanced Japanese international competitiveness in manufactured goods. Here MITI has played a crucial initiating role. Moreover, Japanese policies have been coherent, with private and public banks and private and public research groups working in a coordinated manner towards a set of common goals.[24] This reflects the situation inside major Japanese companies. A very strong case can be made to suggest that direct and coherent policies, as pursued in Japan, are inherently more efficient than indirect and largely fragmented policies of the kind pursued in the United States.

What the above discussion does indicate is that innovative policy making is far from being simply an economic and technological process. Rather, to a very great extent it is a political and cultural process, and policies and practices that succeed in one country may stand little chance of success in other, politically and socially quite different countries. Nevertheless, given the current rapid rate of technological change in a number of key areas, and given the

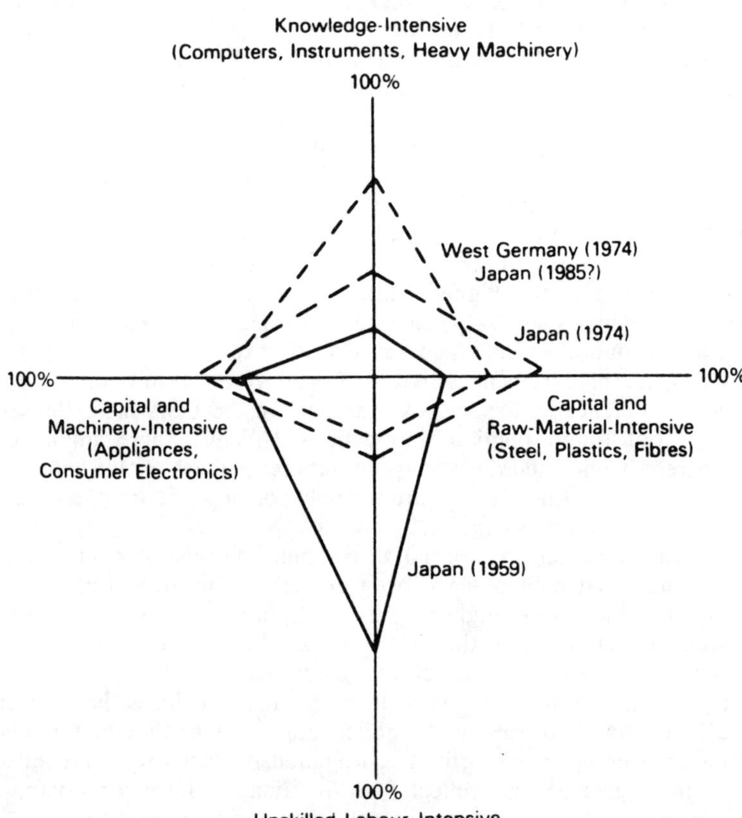

Figure 1　　The evolution of industrial structure

crucial role of technology in economic growth and in international competitiveness, it seems reasonable to suggest that national innovation policies, of whatever sort, are necessary. In other words, it has become an indispensable task of governments to stimulate increased national rates of market-oriented technological change.

Notes

1. R. Rothwell, "The commercialization of university research", *Physics in Technology*, vol.13, no.6, November 1982.
2. S.A. Merrill, "Knowledge and politics in innovation policy design", *Policy Studies Review*, vol.3, no.3-4, 1984.
3. R. Rothwell and W. Zegveld, *Innovation and the Small and Medium Sized Firm*, London, Frances Pinter, 1982.
4. R. Rothwell and W. Zegveld, *Reindustrialization and Technology*, London, Longman, 1985.
5. R. Rothwell and W. Zegveld, *Industrial Innovation and Public Policy*, London, Frances Pinter, 1981.
6. R. Rothwell, *Government Regulations and Industrial Innovation*, Report to the Six Countries Programme on Innovation, Six Countries Secretariat, Delft, The Netherlands, 1979.
7. As mentioned earlier, however, in the case of Great Britain the government has recently become more selective, having designated information technology as a main priority area. Nevertheless, government rhetoric continues to emphasize a "hands off" market forces approach.
8. Rothwell and Zegveld, *Reindustrialization and Technology*.
9. D. Roessner, "Prospects for a national innovation policy in the United States", in this book.
10. R. Dahl, *A Preface to Democratic Theory*, University of Chicago Press, 1958.
11. R. Rothwell, "The role of small firms in the emergence of new technologies", *Omega*, vol.12, no.1, January 1984.
12. G. Dosi, *Technical Change and Industrial Transformation: The Theory and Application to the Semiconductor Industry"*, London, Macmillan, 1984.
13. D. Mowery and N. Rosenberg, "Government policy, technical change, and industrial structure: the US and Japanese aircraft industries 1945-83", in this book.
14. M. Teubel and E. Steinmuller, *Government Policy and Economic Growth*, Research Paper 153, The Maurice Falk Institute for Economic Research in Israel, Mount Scopus, Jerusalem, 1983.
15. R. Nelson (ed.), *Government and Technical Progress*, London, Pergamon Press, 1982.
16. Rothwell and Zegveld, *Reindustrialization and Technology*.
17. Ibid.
18. Ibid.
19. *Science Resources Newsletter*, no.7, Science and Technology Indicators Unit, OECD, Paris, 1983.
20. "Pentagon Inc.", *Frontline America*, Channel 4 Television UK, 10 August 1983.
21. R.M. Colton, T.M. Ryan and D. Senick, "National Science Foundation experiences in stimulating industrial innovation", in this book.
22. G. Allen, "Industrial policy and innovation in Japan", in C. Carter (ed.), *Industrial Policy and Innovation*, London, Heinemann, 1981.

23. This contrasts markedly with Britain where the results of defence R&D become "locked in" to the defence equipment divisions of a small number of principal contractors. J. Maddock, *Civilian Exploitation of Defence Technology*, Report to the Electronics EDC, NEDC, London, February 1983.
24. M.J. Peck and A. Goto, "Technology and economic growth: the case of Japan", *Research Policy*, 10, pp.222-43.

The author wishes to acknowledge the financial support of the Leverhulme Trust Fund *during the preparation of this paper.*